Riccardo Mattei

Radio Science Data Analysis in "Venus Express" and "Rosetta"

AF122654

Riccardo Mattei

Radio Science Data Analysis in "Venus Express" and "Rosetta"

Digital signal processing techniques applied to the reduction of Radio Occultation data

Südwestdeutscher Verlag für Hochschulschriften

Impressum/Imprint (nur für Deutschland/only for Germany)
Bibliografische Information der Deutschen Nationalbibliothek: Die Deutsche Nationalbibliothek verzeichnet diese Publikation in der Deutschen Nationalbibliografie; detaillierte bibliografische Daten sind im Internet über http://dnb.d-nb.de abrufbar.
Alle in diesem Buch genannten Marken und Produktnamen unterliegen warenzeichen-, marken- oder patentrechtlichem Schutz bzw. sind Warenzeichen oder eingetragene Warenzeichen der jeweiligen Inhaber. Die Wiedergabe von Marken, Produktnamen, Gebrauchsnamen, Handelsnamen, Warenbezeichnungen u.s.w. in diesem Werk berechtigt auch ohne besondere Kennzeichnung nicht zu der Annahme, dass solche Namen im Sinne der Warenzeichen- und Markenschutzgesetzgebung als frei zu betrachten wären und daher von jedermann benutzt werden dürften.

Verlag: Südwestdeutscher Verlag für Hochschulschriften GmbH & Co. KG
Dudweiler Landstr. 99, 66123 Saarbrücken, Deutschland
Telefon +49 681 37 20 271-1, Telefax +49 681 37 20 271-0
Email: info@svh-verlag.de

Approved by: München, UniBW, Diss., 2011

Herstellung in Deutschland:
Schaltungsdienst Lange o.H.G., Berlin
Books on Demand GmbH, Norderstedt
Reha GmbH, Saarbrücken
Amazon Distribution GmbH, Leipzig
ISBN: 978-3-8381-2843-6

Imprint (only for USA, GB)
Bibliographic information published by the Deutsche Nationalbibliothek: The Deutsche Nationalbibliothek lists this publication in the Deutsche Nationalbibliografie; detailed bibliographic data are available in the Internet at http://dnb.d-nb.de.
Any brand names and product names mentioned in this book are subject to trademark, brand or patent protection and are trademarks or registered trademarks of their respective holders. The use of brand names, product names, common names, trade names, product descriptions etc. even without a particular marking in this works is in no way to be construed to mean that such names may be regarded as unrestricted in respect of trademark and brand protection legislation and could thus be used by anyone.

Publisher: Südwestdeutscher Verlag für Hochschulschriften GmbH & Co. KG
Dudweiler Landstr. 99, 66123 Saarbrücken, Germany
Phone +49 681 37 20 271-1, Fax +49 681 37 20 271-0
Email: info@svh-verlag.de

Printed in the U.S.A.
Printed in the U.K. by (see last page)
ISBN: 978-3-8381-2843-6

Copyright © 2011 by the author and Südwestdeutscher Verlag für Hochschulschriften GmbH & Co. KG and licensors
All rights reserved. Saarbrücken 2011

To my parents, who taught me Love

To my brother, soul mate and friend

*To Emil, and to all those who, like him,
love the stars*

Acknowledgments

This work came into existence during my activity as project engineer for the Radio Science investigations conducted in the framework of the ESA missions "Venus Express" and "Rosetta".

My heartfelt thanks go to Prof. Dr. Bernd Häusler, Principal Investigator of the Venus Radio Science experiment and Director of the Institute of Space Technology of the University of the German Armed Forces in Munich till 2009. His competent and experienced advice made this work possible. His kind and generous attitude lightened the path.

Deep gratitude ties me to my friend and colleague Dr. Stefan Remus of the European Space Astronomy Centre (ESA/ESAC) at Villanueva de la Cañada (Spain), with whom I share my passion for electronics and signals. His genuine and tenacious enthusiasm accompanied the progress of this work.

I kindly thank Dr. Martin Pätzold, Principal Investigator of the Rosetta Radio Science experiment, for the proficient counsel, and Dr. Silvia Tellmann of the Division of Planetary Science at the "Rheinisches Institut für Umweltforschung" of the University of Cologne for the kind and fruitful cooperation.

The valuable suggestions I received from the ample competence and expertise of Prof. G. Leonard Tyler and Dr. Richard A. Simpson of the Department of Electrical Engineering of the Stanford University are gratefully acknowledged.

I wish to express my gratitude to Dr. Gerhard Staude of the Department of Electronic and Information Technology of the University of the German Armed Forces for the stimulating and productive conversations.

My sincere thanks to Ms. Cornelia Freudenberg of the Library of the University of the German Armed Forces, who immensely eased the bibliographic researches with grace and humour.

Finally, my special thanks go to all those who -relatives, friends, acquaintance, or strangers- offered me with the precious gift of a smile.

Contents

Acknowledgments	iii
List of Abbreviations	x
List of Symbols	xiii
Preface	xix

1 The ESA Missions "Venus Express" and "Rosetta" — 1
   1.1 The "Venus Express" mission 1
       1.1.1 Mission overview: mission design and operations 2
       1.1.2 Mission Planning 4
       1.1.3 VEX Spacecraft technical constraints 4
   1.2 The "Rosetta" mission 6
       1.2.1 Mission overview: mission design and operations 8
   1.3 Communications, Ground Segment, and
       Spacecraft Communications-Subsystem 9
       1.3.1 Link-Budget 14

**2 The Venus Radio Science Experiment "VeRa" and
the Rosetta Radio Science Investigation "RSI"** — 18
   2.1 Fundamentals of Radio Science 18
   2.2 Instrumentation and operational modes 19
   2.3 Brief description of the Radio Science Experiments 22
       2.3.1 The "Gravity" Experiment (GRA) 25
       2.3.2 The "Solar Corona" Experiment (SCO) 26

3	**The Radio Science Simulator (RSS)**		**30**
	3.1	Concept & lay-out of the Radio Science Simulator	30
		3.1.1 Modules of the Radio Science Simulator	31
	3.2	Time Bases and Reference Systems	33
		3.2.1 Time Bases	34
		3.2.2 Reference Systems	37
		3.2.3 Reference frame and time basis of input data	39
4	**The Bistatic Radar Experiment**		**40**
	4.1	The concept of the "Bistatic Radar" experiment	40
	4.2	Experiment geometry	42
	4.3	Power frequency spectrum of echo signal	44
		4.3.1 Doppler-shift	45
		4.3.2 Spectral broadening	45
		4.3.3 Echo strength	46
	4.4	Estimate of the dielectric constant of the planetary surface	48
		4.4.1 Reflection coefficients	50
	4.5	Computation of expected Doppler frequency shift of echo signal	54
		4.5.1 Implementation of the "BSR Doppler Module" in the RSS . .	55
5	**The Radio Occultation Method**		**62**
	5.1	The "Occultation" Experiment	62
	5.2	Measurement principle	64
		5.2.1 Actual optical path	69
		5.2.2 Atmospheric Profiles	71
		5.2.3 Signal Attenuation: Absorption and Defocusing-Loss	74
		5.2.4 Spatial Resolution	77
	5.3	Data kind and data acquisition	79
6	**Elements of Signal Processing**		**85**
	6.1	Analog signals	86
		6.1.1 Finite-Energy Signals and Finite-Power Signals	86
		6.1.2 The Fourier Series	87
		6.1.3 The Fourier Transform	94

	6.1.4	Cross- and Autocorrelation; the Wiener-Kintchine Theorem	100
	6.1.5	Linear Time Invariant Systems	102
	6.1.6	Modulation	104
	6.1.7	The analytic signal and the Hilbert Transform	105
6.2	Discret-Time Signals[1]		110
	6.2.1	Sampling theorem	110
	6.2.2	The Fourier Transform of discrete-time signals	113
	6.2.3	The Discrete Fourier Transform (DFT)	114
	6.2.4	The "z-transform"	117
	6.2.5	LTI Systems for Discrete-Time Signals	119
	6.2.6	Numeric filters	120
6.3	Elements of Spectral Estimation[2]		128
	6.3.1	Estimation Theory	130
	6.3.2	Classical Spectral Estimation	131

7 "Open Loop" Data Processing Software — 137

7.1	Basic principle		137
	7.1.1	Implementation of predicted expected frequency shift in the OL software	143
	7.1.2	Single iteration steps: signal frequency evaluation, polynomial fit and storage of coefficients	150
	7.1.3	Computation of last frequency residual and received signal power from time samples and construction of the total frequency shift	151
	7.1.4	FFT-based frequency estimation	155
	7.1.5	Design of numeric low-pass filters	158
7.2	Multipath propagation		161
	7.2.1	The "Multi-track" routine	163
	7.2.2	Processing techniques for multipath-affected signals	170
	7.2.3	The Wigner-Ville Transform	179

8 Error Analysis — 189

8.1	Statistical Error	190

---

[1] An extensive treatment of this subject can be found in [71].
[2] An extensive treatment of this subject can be found in [58]. See also [71].

	8.1.1	Estimation of the receiver thermal noise 190
	8.1.2	Uncertainties in the calculated frequency- and power profiles . 195
	8.1.3	USO Phase Noise 204
	8.1.4	Quantization noise 206
	8.1.5	Fluctuation in the propagation path 207
8.2	Systematic Errors 207	
	8.2.1	Inaccuracy in the evaluation of the effects of the Earth ionosphere and troposphere 208
	8.2.2	Mis-pointing of the S/C antenna 208
	8.2.3	Oscillator drift (aging) 209
	8.2.4	Inaccuracy in the reconstructed orbit data and in the S/C attitude 209
	8.2.5	Execution time offset 209
	8.2.6	Finite precision representation 210
	8.2.7	Computational Errors 211
	8.2.8	Algorithm 211

9 Summary   **212**

A Analog PM and FM modulations[3]   **216**

B The Signal-to-Noise Ratio (SNR)   **220**

C The Allan Variance   **227**

D The Ultra Stable Oscillator (USO)   **230**
    D.1 The USO design 230
    D.2 The calibration measurements 232

E Refraction and reflection of plane waves[4]   **235**

F Probability, Random Variables,
and Stochastic Processes[5]   **239**
    F.1 Probability 239

---

[3]An extensive treatment of this subject can be found in [17].
[4]An extensive treatment of this subject can be found in [99].
[5]An extensive treatment of this subject can be found in [74].

F.2	The random variable	240
	F.2.1 Moments of a random variable	240
	F.2.2 Joint moments	241
	F.2.3 Uncorrelated, Orthogonal, Independent random variables . . .	242
F.3	Stochastic Processes	243
	F.3.1 Time averages, ensemble averages and ergodicity	246

G The Wigner Distribution[6]   249

   G.1 Analog signals 250
   G.2 Discrete signals 254

H Quantization Error   257

---

[6]An extensive treatment of this subject can be found in [12], [13], [14], and [15].

List of Abbreviations

ACF	....	Auto-Correlation Function
ADC	....	Analog-to-Digital Conversion
AGC	....	Automatic Gain Control
AWGN	....	Additive White Gaussian Noise
BP	....	Backward- Propagation
BSR	....	Bistatic Radar
BVA	....	Bote Vieillage Ameliore
BW	....	Band Width
CL	....	Closed Loop
CME	....	Coronal Mass Ejection
DC	....	Direct Current
DDID	....	Data Delivery Interface Document
D/L	....	Down-Link
DoY	....	Day-of-Year
DSN	....	Deep Space Network
EGSE	....	Electrical Ground Support Equipment
ESA	....	European Space Agency
ESOC	....	European Space Operation Centre
ESU	....	External Storage Unit
FD	....	Flight Dynamics
FM	....	Frequency Modulation
FOV	....	Field Of View
GPS	....	Global Positioning System
GRA	....	Gravity
G/S	....	Ground Station

HGA	....	High Gain Antenna
IAU	....	International Astronomical Union
ICD	....	Interface Control Document
IEEE	....	Institute of Electrical and Electronics Engineers
IERS	....	International Earth Rotation and Reference Systems Service
IFMS	....	Intermediate Frequency and Modem System
ITRS	....	International Terrestrial Reference System
JPL	....	Jet Propulsion Laboratory
LEO	....	Low Earth Orbit
LO	....	Local Oscillator
LTAN	....	Local Time at Ascending Node
LTI	....	Linear Time Invariant
MCC	....	Mission Control Centre
MCS	....	Mission Control System
MPS	....	Multiple-Phase-Screen
MSE	....	Mean Squared Error
NASA	....	National Aeronautics and Space Administration
OCC	....	Occultation
OL	....	Open Loop
PDF	....	Probability Density Function
PDS	....	Power Density Spectrum
PLL	....	Phase Lock Loop
PM	....	Phase Modulation
RF	....	Radio Frequency
RMS	....	Root Mean Square
RNG	....	Ranging
RO	....	Radio Optic
RS	....	Radio Science
RSI	....	Radio Science Investigation
RSS	....	Radio Science Simulator
SAP	....	Science Activity Plan
SAR	....	Synthetic Aperture Radar
S/C	....	Spacecraft
SCO	....	Solar Corona

SI	....	System of Units, *Systme international d'units*
SNR	....	Signal-to-Noise ratio
SS	....	Sliding Spectra
SW	....	Software
TC	....	Telecommands
TCXO	....	Thermal Controlled Crystal Oscillator
TEC	....	Total Electron Content
TM	....	Telemetry
TRSP	....	Transponder
TT&C	....	Tracking, Telemetry and Commands
TWTA	....	Traveling Wave Tube Amplifier
U/L	....	Up-Link
USO	....	Ultra Stable Oscillator
VeRa	....	Venus Radio Science
VEX	....	Venus Express
VMOC	....	Venus Mission Operation Center
VSOC	....	Venus Science Operation Center

List of Symbols

α	....	bending angle (rad)
β	....	modulation index [-]
Γ	....	power reflection coefficient [-]
Δ	....	value of the upper extreme of the quantization scale [-]
Δt	....	sampling time $[s]$
ϵ	....	permittivity $[F \cdot m^{-1}]$
ϵ_0	....	electric constant $[F \cdot m^{-1}]$
ε_A	....	amplitude uncertainty [-]
ε_P	....	power uncertainty [-]
ϵ_r	....	dielectric constant (or relative permittivity) [-]
η	....	antenna efficiency [-]
η	....	characteristic impedance $[\Omega]$
θ_B	....	Brewster angle (rad)
θ_i	....	incidence angle (rad)
θ_r	....	reflection angle (rad)
θ_t	....	transmission angle (rad)
λ	....	wave-length of an electromagnetic field $[m]$
μ	....	permeability $[H \cdot m^{-1}]$
μ	....	refractivity [-]
μ_0	....	magnetic constant $[H \cdot m^{-1}]$
μ_k	....	generic central k^{th} moment of a random variable [-]
μ_{kr}	....	generic central $(k+r)^{th}$- order moment of two random variables [-]
μ_r	....	relative permeability [-]
ξ	....	rms surface slope [-]

ξ_j	....	value of the lower extreme of the generic j^{th} quantization interval [-]
σ	....	radar cross section (RCS) $[m^2]$
σ_ϕ	....	phase uncertainty (rad)
σ_{AV}	....	frequency instability $[Hz]$
σ_f	....	frequency uncertainty $[Hz]$
σ_q	....	quantization error [-]
σ_v	....	velocity error $[m \cdot s^{-1}]$
σ_y	....	Allan deviation [-]
σ^2	....	variance of a random variable [-]
τ	....	defocusing loss [-]
τ	....	group delay $[s]$
τ	....	optical depth [-]
Υ	....	power transmission coefficient [-]
Φ	....	gravitational potential $[m^2 \cdot s^{-2}]$
ϕ_a	....	azimuth angle (rad)
ϕ_i	....	incidence angle (rad)
ϕ_s	....	scattering angle (rad)
Ω_f	....	first-order moment of the Wigner Distribution w.r.t. the frequency $[Hz]$
ω_p	....	plasma frequency $[s^{-1}]$
a	....	impact parameter $[m]$
A_e	....	antenna effective aperture $[m^2]$
A_{eff}	....	effective area $[m^2]$
AU	....	Astronomical Unit $[m]$
B_c	....	receiver bandwidth for carrier $[Hz]$
B_{data}	....	receiver bandwidth for data $[Hz]$
BW	....	bandwidth $[Hz]$
C	....	autocovariance [-]
C	....	signal power $[W]$
c	....	speed of light in vacuum $[m \cdot s^{-1}]$
c_k	....	Fourier coefficients [-]
dB	....	decibel [-]
e	....	electron charge [C]
F	....	receiver noise figure [-]

f_c	....	carrier frequency $[Hz]$
f_s	....	sampling frequency $[s^{-1}]$
G	....	gravitational constant $[m^3 \cdot kg^{-1} \cdot s^{-2}]$
G_R	....	reception antenna gain [-]
G_T	....	transmission antenna gain [-]
g	....	acceleration of gravity $[m \cdot s^{-2}]$
I_j	....	generic j^{th} quantization interval [-]
J_n	....	Bessel's function of the first kind of order n [-]
K	....	Boltzmann's constant $[W \cdot K^{-1} \cdot Hz^{-1}]$
k	....	wave number $[m^{-1}]$
L_{FS}	....	free-space loss [-]
$\bar{m}$	....	mean molecular mass $[kg]$
m_e	....	electron mass $[kg]$
m_k	....	generic k^{th} moment of a random variable [-]
m_{kr}	....	generic $(k+r)^{th}$- order moment of two random variables [-]
N	....	noise power $[W]$
N_0	....	noise spectral power density $[W \cdot Hz^{-1}]$
N_e	....	electron number density $[m^{-3}]$
n	....	AWGN noise process $[V]$
n	....	refractive index [-]
p	....	pressure $[N \cdot m^{-2}]$
P_{LCP}	....	*left circular polarization* received power $[W]$
P_{RCP}	....	*right circular polarization* received power $[W]$
P_R	....	received power $[W]$
P_T	....	transmitted power $[W]$
p_x	....	probability density function of the r.v. x [-]
q	....	width of quantization interval [-]
q_j	....	value assigned to the the generic j^{th} quantization interval [-]
R	....	(auto)correlation function [-]
R_H	....	horizontal reflection coefficient [-]
R_{OC}	....	*opposite circulation sense* reflection coefficient [-]
R_{SC}	....	*same circulation sense* reflection coefficient [-]
R_V	....	vertical reflection coefficient [-]
r	....	correlation coefficient of two random variables [-]
r_{eff}	....	radius of the effective area $[m]$

r$_H$	....	radius of the first Fresnel zone (horizontal) $[m]$
r$_V$	....	radius of the first Fresnel zone (vertical) $[m]$
S	....	signal power $[W]$
SNR	....	signal-to-noise ratio [-]
SNR$_0$	....	signal-to-noise density ratio $[Hz]$
T	....	observation period $[s]$
T	....	absolute temperature $[K]$
T$_A$	....	equivalent antenna temperature $[K]$
T$_{in}$	....	input temperature $[K]$
T$_{sys}$	....	system temperature $[K]$

Preface

Since their discovery and description in the nineteenth century, electromagnetic waves have become an essential instrument for the advancement of the human race. Among the prominent scientists who contributed with pioneer work to the classical electromagnetic theory, we would like to mention here the names of H. C. Ørsted, J. Henry, A. Ampere, M. Faraday, J. C. Maxwell, H. R. Hertz. Since the first exciting experiments at the beginning of the last century,[7] electromagnetic waves has definitely established as a mean of communication over long distances. These early experiments led to the discovery that the Earth ionosphere can reflect incident electromagnetic waves, depending on their wavelength. The first radio wave measurement of ionospheric reflecting layers dates back to 1924 ([101]). Also the possibility to detect distant reflecting objects by means of electromagnetic waves was already known by that time, but it was only during the World War II that this discovery was transformed into an operative defense system, which was given the name of *RADAR*: **RA**dio **D**etection **A**nd **R**anging. We regard this circumstance as the birth of the *Remote Sensing*: the science and technology intended to acquire information about distant objects by means of opportune sensing devices which gather electromagnetic energy emitted or reflected by the investigated object. The possibility to have the sensing device in the vicinity of the physical object to be analyzed and to exploit radio waves[8] to transmit the attained information to the remote experimenters plays a central role in space exploration, as ground-based observation at Earth are severely limited by the distance of the investigated objects and –depending on the kind of the observation– by the presence of the Earth atmosphere. Spacecraft missions offer the opportunity to conduct observation from a close distance in addition to *in situ* measurements, where a probe is released by the

---

[7]We cite here among others the names of G. Marconi, A. Meucci, A. G. Bell.
[8]Electromagnetic waves whose frequency is suitable for transmission/reception.

spacecraft to directly measure the medium of interest.

Remote sensing techniques are grouped into two categories which are known as *active* remote sensing, and *passive* remote sensing. In the active remote sensing the investigated object is excited through electromagnetic radiation emitted by the sensing equipment (i.e. a radar system). The reflected radiation is then collected by the same device (i.e. *mono-static* radar) or by a different one (i.e. *bi-static* radar) and stored for analysis. In passive remote sensing the probed medium is not artificially excited: the probing device gathers electromagnetic energy which is either spontaneously emitted by the observed object or reflected, being the source of radiation of natural origin (i.e. the sun, a star, etc.). An example of passive remote sensing are radiometric measurements. In this case the use of different wavelengths and observation times help to separate emitted and reflected radiation coming from the target.

A particular kind of remote sensing, which can be classified neither as passive, nor as active (although emission of radio-waves by the probing equipment is implemented) is *Radio Science* (RS)[9]. The Radio Science technique is based on the interaction between electromagnetic waves and physical media which may be encountered by a radio signal during its propagation from the transmitting- to the receiving antenna. The media lying on the radio signal path induce variations on some signal parameters –set constant at transmission– which can be relate to some physical properties of the crossed media. Therefore, the practicability of remote investigation of physical media by means of radio science depends on the possibility to retrieve time-dependent variations of some parameters of the received radio signal. If proper reception is guaranteed, it is possible to extract the looked-for information by means of opportune signal processing techniques.

Among the Radio Science observation techniques, "occultation" experiments exploit a special observation geometry in which the planet lies on the radio path between spacecraft and Earth in order to let the RF carrier signal transmitted by the satellite propagate through planetary ionosphere and atmosphere. Due to the interaction with the interposed media, signal parameters such as frequency, amplitude, polarization, group delay undergo modifications which allow investigators to

---

[9]One of the different observation techniques of radio science, the "bistatic radar" experiment, can be indeed regarded as an active remote sensing technique.

characterize the media itself.

Occultation measurements are carried out in the frame of the Radio Science experiment on board the *"Venus Express"* (VEX) mission, which is part of the solar system exploration program of the European Space Agency (ESA). Particularly severe are the conditions imposed by the thick Venus atmosphere on the occultation experiment: the deeper atmospheric layers encountered by the RF carrier signal transmitted by the VEX spacecraft during the sounding cause remarkable changes and change rates of the signal parameters. Degradation of the Signal-to-Noise Ratio (SNR) can exceed $50\,dB$, depending on the observation geometry; the frequency shift caused by the interposed medium can reach up to $60\,kHz$, with frequency shift rate exceeding $2\,kHz/s$. Under such conditions, PLL-based receivers, which constitute the "Closed Loop" receiving system (CL), are very soon driven in the out-of-lock status and data acquisition stops. The implementation of the so-called "Open Loop" (OL) receiving technique has revealed of enormous advantage when dealing with critical values of signal attenuation and dynamics, as it eliminates real-time tracking of the signal, thus overcoming the problem of conflicting requirement on the PLL-receiver bandwidth. (High signal dynamics require large reception bandwidth, whereas degraded SNR imposes narrow-band reception). The incoming voltage-signal is sampled at proper rate, dictated by the Nyquist criterion, which safeguards the high dynamics impressed onto the signal by the deep layers of the planetary atmosphere. The possibility to recover the weak S/C signal sunk in the receiver thermal noise lies in digital post-processing techniques. For this reason, Venus Express is the first Venus mission to implement an *Ultra-Stable-Oscillator*, the USO, whose purpose is to provide the Radio Science investigation with a highly stable frequency reference driving the probing RF carrier signal. This implies the possibility in the post-processing phase to reduce the noise bandwidth below $1\,Hz$, which is the limit at which the contribution of the phase noise becomes appreciable.

This work is concerned with algorithms for the analysis of radio science data, particularly with the digital signal processing techniques developed and applied at the Institute of Space Technology at the German Armed Force University, "Universität der Bundeswehr", in Neubiberg (Munich) to the *Open Loop* data arising from radio science occultation measurements at Venus.

The developed routines will also be suitable for data analysis of the Radio Science Investigation (RSI) carried out within the *"Rosetta"* mission, ESA exploration project of high scientific relevance, currently in *cruise* toward its target, the comet *"67P/Churyumov-Gerasimenko"*. Like Venus Express, the Rosetta spacecraft is equipped with an USO which makes it possible to adopt -for the first time in a cometary mission- the "One-Way" transmission mode[10] for Radio Science experiments.

Besides the OL data evaluation software, computational routines were developed for the radio science bistatic radar experiment, which is performed on both Venus Express and Rosetta missions. Target of the programmed software tool is the calculation of the expected Doppler frequency shift of the echo signal from the surface of the investigated planetary body. The availability of a such a prediction is indispensable to the proper setting of the ground receiving system during the performance of the experiment and also of great advantage in the post-processing of the data.

After a brief general insight on missions and spacecraft, the radio science (RS) experimental techniques are discussed, in particular the occultation experiment (OCC) and the bistatic radar experiment (BSR). The on-board radio science instrumentation and ground equipment are described along with the different receiving techniques and data kinds. The inherent advantages of the open loop recording strategy toward the closed loop technique are discussed. Additional benefits brought by the OL recording technique in multipath environments are addressed. The Open Loop software, the package which encompasses all processing routines, is presented along with preliminary results and error analysis.

---

[10]Transmission mode in which the RF carrier signal is sent from the spacecraft to the ground station (see Ch. 2).

Chapter 1

The ESA Missions "Venus Express" and "Rosetta"

1.1 The "Venus Express" mission

Orbiting the Sun at a medium distance of 0.7 AU in a quasi-circular orbit of period 224.7 days (sidereal), the planet Venus is the Earth's next neighbor. With a mean radius length of 6051.8 km, corresponding to 0.949 Earth radii and a mass of $4.869 \times 10^{24}\,kg$ (0.815 times the mass of the Earth), Venus shows at first sight a certain similarity with our planet ([10]). But early studies of the Earth's next neighbor, based both on ground observations and spacecraft missions (Venera, Pioneer-Venus and Vega, 1962-1985), had reveled a unique scenario: Venus is surrounded by a thick atmosphere consisting mainly of CO_2 ($\sim$96.5%) and N_2 ($\sim$3.5%), shrouded by an inhomogeneous sulfuric acid clouds layer extending for $\sim$20 km in the middle atmosphere. The dense gases cause a pressure at ground exceeding 90 bar, whereas the greenhouse effect determines a surface temperature as high as 735 K. Starting from the cloud tops, with velocities decreasing from $\sim$100 m/s to near zero at the surface, strong winds cause this portion of the Venus atmosphere to rotate around the planet in about four-five terrestrial days, much faster than the planet itself (Venus rotation period (sidereal day): $\sim$243 terrestrial days, retrograde), a phenomenon known as *"superrotation"* which is not yet completely understood ([46], [93]).

The VEX mission science objectives have been grouped in seven categories: ([93])

- atmospheric structure;

- atmospheric dynamics;

- atmospheric composition and chemistry;

- cloud layer and hazes;

- energy balance and greenhouse effect;

- plasma environment and escape processes;

- surface properties and geology.

The synergy of the investigation instruments on-board Venus Express, which compose the so called "payload", guarantees the coverage of all relevant issues. In particular the Radio Science experiment makes contributions to different topics: structure and dynamics of the atmosphere, plasma and escape processes, clouds and surface.

1.1.1 Mission overview: mission design and operations

The conception of the "Venus Express" (VEX) mission was born from the idea of the European Space Agency (ESA) to exploit know-how, technology, and hardware developed for the missions "Rosetta" and "Mars Express". The proposal which won the competition was to re-use the Mars Express bus with some opportune changes, needed mainly because of the smaller distance of Venus from the Sun. The expertize gained in the Rosetta and Mars Express missions, along with the possibility to employ the same industrial and scientific teams, made it possible to ESA to prepare the mission in less than four years ([93]). The Venus Express spacecraft was launched on November 9, 2005 from the Baikonur cosmodrome, Kazakhstan, by a Soyuz-Fregat rocket and successfully inserted in orbit around Venus on April 11, 2006. The mission is based on a highly elliptical polar orbit (semi-major axis a = 39494 km, eccentricity e = 0.84, height of pericenter maintained by thruster firings in the range between 180 km and 250 km). The orbit period is 24 h. The operations are directed by the ESA Mission Control Centre (MCC) located at the European Space Operation Centre (ESOC) in Darmstadt, Germany, via the ESA ground stations in Cebreros (Spain) and in New Norcia (Australia). Cebreros is employed for communication tasks, while the New Norcia ground station supports the Radio Science investigation. The location of both stations ensures radio coverage during almost the complete VEX spacecraft orbit ([68], [93]).

The VEX payload consists of seven instruments designed and developed by research institutes and space industries of participating countries (see fig.1.1). These are: three spectrometers: *VIRTIS*, an imaging and high-resolution spectrometer in the spectral range from visible through thermal infrared ([78], [96]), *PFS*, high-resolution infrared spectrometer ([25], [96]), and *SPICAV*, a high-resolution UV and near-IR spectrometer for stellar and solar occultation and nadir observations ([7], [96]); a camera, *VMC* , operating in the visible and near-IR range ([65], [96]), a plasma analyzer, *ASPERA* ([96]), for in situ investigation of plasma and neutral energetic atoms around the planet ([5], [96]), a magnetometer, *MAG* ([96]), for magnetic field measurements ([105], [96]). The radio science experiment *VeRa* is conducted by means of an ultra-stable oscillator (USO), which provides the investigators with the needed stability of the frequency source for the purpose of studying the fine structure of the atmosphere and ionosphere, investigating surface properties and gravity anomaly and performing solar corona sounding ([45], [46], [96]).

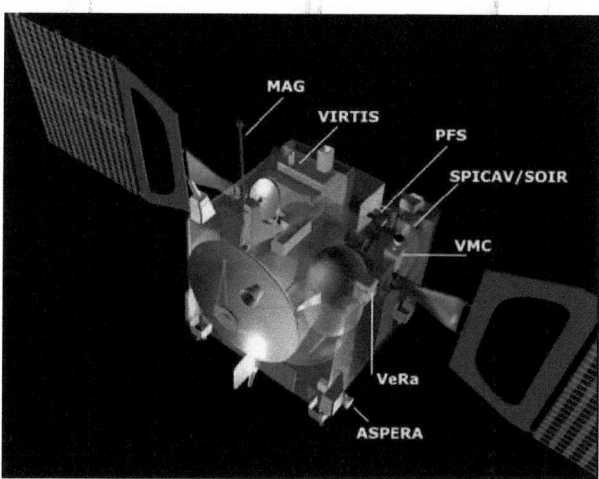

Figure 1.1: *Payload accomodation on-board VEX.* ([93])

1.1.2 Mission Planning

Task of the mission planning process is to transform the top-level mission science targets into operative procedures by taking into account both, scientific goals and operational constraints. To this purpose, the main science topics are decomposed into sub-themes which, in turn, lead to detailed scientific objectives. These, along with operational characteristics such as trajectory, environment and spacecraft constraints, define different kinds of possible instrument measurements, the *observations* or *science cases* (see fig. 1.2). A science case establishes the geometry of the observation by defining the satellite position in the orbit during the measurement (i.e. from apocenter, pericenter, ascending branch, etc.), the instruments which shall carry out the observation, and the observation duration (see fig. 1.3). This high-level mission planning is ruled by the *Science Activity Plan* (SAP) and is part of the *long-term planning*, which started two years before the launch. Routine operations are planned monthly and scheduled in *medium-term* plans (MTP) ([96]).

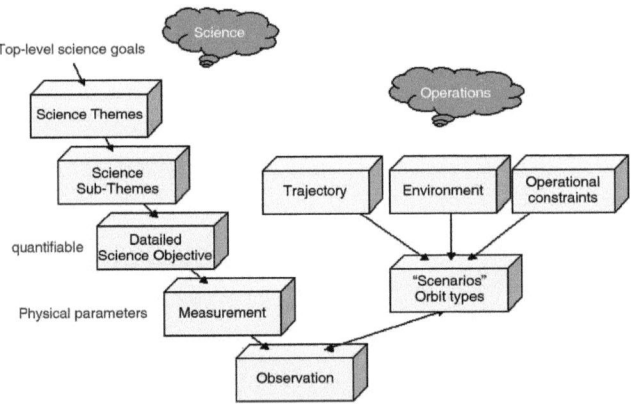

Figure 1.2: *VEX high-level planning scheme* ([96]).

1.1.3 VEX Spacecraft technical constraints

Technical constraints of the S/C have to be respected when planning observations, first of all thermal constraints, as certain S/C panels are designed for heat exchange or host payload parts sensitive to solar heating. As a result, they must be kept

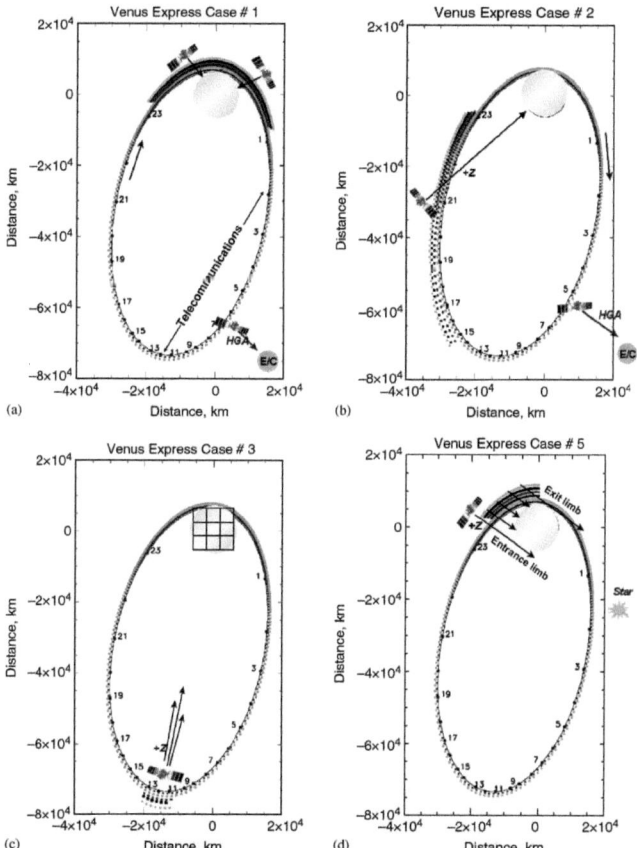

Figure 1.3: *Example of science cases #1, #2, #3 and #5. Thicks along the orbit mark the time in hours past pericenter; solid lines and dots mark different kind of payload activity. (For a detailed description of the VEX science cases see [96])*

in the shadow throughout the mission or are allowed to be directly illuminated by solar radiation only for a certain amount of time under certain angles. The only panel which can be fully illuminated by the Sun for long time durations is the $+X$ panel (reference system is defined in fig. 1.4), where the large dish of the main high-gain antenna (HGA1) acts as a sun-shield. The minimum incidence angle allowed for the incoming radiation on the $+Z$ panel is $10\,°$ because of the presence of sensible optical instruments. The $\pm Y$ panels (attachment points of the solar arrays) can be exposed under an incidence angle $\geqslant 80\,°$, but for no longer than $1h$ (see fig. 1.4), whereas $-Z$ and $-X$ panels can be exposed only within very limited profiles (sun aspect angle on the spacecraft wall vs. time, [96], [19]). Furthermore, mechanical constraints allow a maximum S/C angular rate of $0.15 deg/s$ ([68], [96]). All constraints have been implemented by the Radio Science experiment team in the algorithms which compute the S/C attitude for different radio science observational modes. Depending on the particular experiment to be executed and on the relative positions of the involved celestial bodies, S/C, and Sun, the mentioned constraints pose limitations which sometimes lead to the impossibility to perform the desired measurement.

1.2 The "Rosetta" mission

The Rosetta mission is named after the Rosetta stone, the basaltic stele found by French soldiers in 1799 near the village of Rashid (Rosetta), at the Nile's delta. As the stone was carved both with hieroglyphs and ancient Greek text, it was possible for the English physician and physicist Thomas Young and the French scholar Jean François Champollion to decipher the ancient Egyptian writing, thus opening the access door to one of the most interesting cultures of the antiquity. The analogy with the Rosetta mission lies in the consideration that comets travelling the solar system are leftovers of the vast swarm of asteroids and comets which surrounded the Sun at the time when planets were not yet formed. As comets have not undergone the geological processes which are typical for large bodies (planets), they can reveal interesting clues on the origin of the solar system, thus offering the scientists with the "deciphering key" for the origin of the solar system, as the Rosetta stone did for the hieroglyphs ([21]).

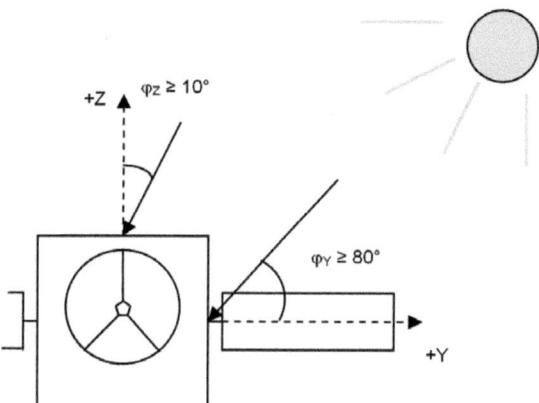

Figure 1.4: *Simple representation of some thermal constraints imposed on the VEX spacecraft (+Y, and +Z panels). The same protection policy adopted for the +Y panel is also valid for the −Y panel, whereas the −Z, and −X can be exposed only within very limited profiles (sun aspect angle on the spacecraft wall vs. time, [68], [96], [19]).*

Spacecraft-based observations and *in situ* measurements of comets performed by precedent cometary missions revealed comet activity which is rather localized at the nucleus surface. Only minor parts are active and the activity is highly constant on short time scales. Sometime dramatic outbursts with increase of activity occur, but their origin is still unclear. Many physicochemical processes, like (photo)chemical reactions, and interaction with solar wind alter the material originally present in the nucleus. All the present results indicate that cometary nuclei are unique, have their own history, and are not entirely pristine ([28]).

In 2014 the Rosetta spacecraft will encounter comet *67P/Churyumov-Gerasimenko* close to its aphelion and will study the physical and chemical properties of the nucleus, the evolution of the coma during the comet's approach to the Sun, and the development of the interaction region of the solar wind and the comet for more than one year, till the comet will reach perihelion ([28]).

The Rosetta mission science objectives include: ([28])

- global characterization of the nucleus;

- determination of nucleus' dynamic properties;

- surface morphology and composition;

- determination of chemical, mineralogical, and isotopic compositions of volatiles and refractories in the cometary nucleus;

- determination of physical properties and interrelation of volatiles and refractories in the cometary nucleus;

- studies of the development of cometary activity and the processes in the surface layer of the nucleus and inner coma, that is dust/gas interaction;

- studies of the evolution of the interaction region of the solar wind and the outgassing comet during perihelion approach.

The Radio Science experiment in the frame of the Rosetta mission is concerned with modeling of gravitational properties of the comet, with the radar probing of the comet surface, the occultation and near-forward scattering by the coma particles and occultation by the coma ionosphere and neutral gas.

1.2.1 Mission overview: mission design and operations

The Rosetta mission, cornerstone in the ESA's long term program "Horizons 2000" and logical follow-up of the very successful ESA mission *Giotto* to comet "1P/Halley", is intended to investigate the origin of our solar system by studying the origin of comets. Originally planned as a comet nucleus sample return mission, it consists of two mission elements, the Rosetta orbiter and the lander "Philae". A cooperation between ESA, several national space agencies, and NASA, the Rosetta mission, being comprised of 25 experiments, is unprecedented in scale.

The original target of the mission was comet *46P/Wirtanen*, but a rocket failure shortly before launch forced ESA to re-schedule the mission start and to re-target Rosetta, now heading for comet *67P/Churyumov-Gerasimenko*. Launched on March 2, 2004 from the French Guyana by an Ariane-5 G+ launch vehicle, the Rosetta spacecraft will travel the distance Sun-Earth five times, during a circuitous 10-year trek, and will pass through the asteroid belt into deep space beyond 5AU solar distance. Four planetary gravity assist manoeuvres (Earth-Mars-Earth-Earth), successfully executed between 2005 and 2009, shall confer the spacecraft the sufficient energy to reach its target. We mention here the successfully Rosetta flyby with the asteroid "Lutetia", on July 10, 2010, which is of great significance for the Radio Science Investigation of the asteroid gravitational properties (see [3]).

After the first *rendezvous* manoeuvre on January 2011, the S/C will enter the "hibernation" mode till early 2014. After a second rendezvous manoeuvre on March 2014, "Near-Nucleus" operations will be started. The "Philae" lander will be delivered on November 2014, close to the start of the "Comet Escort" phase to the perihelion passage on August 2015. End of Nominal Mission: December 31, 2015 ([28]).

Fig. 1.5 shows the Rosetta spacecraft design. Tables 1.1, and 1.2 summarize the payload of the Rosetta orbiter, and of the Rosetta lander, Philae, respectively. (See [28]).

1.3 Communications, Ground Segment, and Spacecraft Communications-Subsystem

Information exchange between the Mission Control Center and the probe is of central importance for spacecraft-based space missions. To this purpose a radio link between the controlling ground station and the spacecraft must be established. In this regard, the main concern are the large distances which are typical for space exploration missions, as the power of the transmitted electromagnetic waves distributes on spherical surfaces centered at the transmitting source. This causes the power of the received radio signal to be inversely proportional to the square of the traveled distance, which implies that only a tiny fraction of the transmitted power will reach the receiving antenna (*free-space loss*, see App. B). On ground, employment of large antenna dishes -up to $35\,m$ in diameter- and cooling of the radio receivers by means of liquid helium make it possible to retrieve the weak S/C radio signals out of the receiver- and background noise. Values of the irradiated electromagnetic power at the G/S as large as $20kW$ guarantee proper reception on the S/C side. Transmission from the G/S to the S/C is indicated as *up-link* (U/L), whereas the reverted communications flow is indicated as *down-link* (D/L).

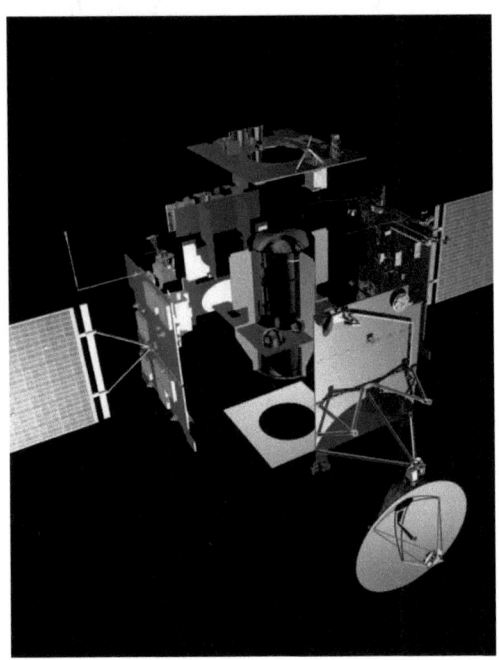

Figure 1.5: *Rosetta spacecraft design. Image: ESA.* ([22]).

The VEX spacecraft is equipped with two high-gain antennas, *HGA1* and *HGA2*, and two omni directional antennas, *LGA1* and *LGA2* (see fig. 1.6). The *HGA1*, with a dish diameter of $1.43\,m$ and two feeds for S- and X-Band, is mounted on the S/C $+X$ wall and serves also as sun shield. The antenna gain at X-Band is $38.6\,dBi$ for the up-link and $39.8\,dBi$ for the down-link. At S-Band the gain is $27.8\,dBi$ and $28.5\,dBi$, respectively for up- and down-link. The *HGA2*, with a dish diameter of $0.2,m$ and only a X-Band feed (gain $21.5\,dBi$ and $22.7\,dBi$, up- and down-link, respectively) is mounted on the top S/C panel ($+Z$ wall) and aligned along the $-X$ axis in order to guarantee communications during Venus inferior conjunction.[1] The two omni-directional antennas, which are operated at S-band as back-up for the

---

[1]When the planet is inside the quadrature positions, the sun shines from behind the spacecraft. Since no solar illumination is allowed on the $-X$ wall (which carries a "cold plate" for the radiometers), the S/C must be flipped.

Instrument Name	Scientific Objectives
ALICE	UV-Spectroscopy ($70\,nm - 205\,nm$)
CONSERT	Radio sounding and nucleus tomography
COSIMA	Dust mass spectrometer
GIADA	Dust velocity and impact momentum measurement
MIDAS	Grain morphology
MIRO	Microwave-Spectroscopy ($1.3\,nm\,and\,0.5\,nm$)
OSIRIS	Multi-color imaging
ROSINA	Neutral gas and ion mass spectroscopy
RPC	Ion composition analyzer
	Ion and electron sensor
	Langmuir probe
	Fluxgate magnetometer
	Mutual impedance probe
	Plasma interface unit
RSI	Radio science experiment
SREM	Standard radiation environment monitor
VIRTIS	VIS and IR mapping spectroscopy ($0.25 - 5\,\mu m$)

Table 1.1: *Payload of the Rosetta orbiter.* ([28]).

communication system, are mounted on the spacecraft $+Z$ wall (top panel) and $-Z$ wall (bottom panel). All antennas are rigidly fixed to the bus. This implies that antenna pointing has to be executed by S/C attitude maneuvers. Two on-board transponders (TRSP#1, and TRSP#2)[2] receive and transmit RF-signals at S- and X-Band with an output power of $5W$ (solid state amplifier) and $65W$ (traveling wave tube amplifier, TWTA), respectively ([6]). A system of switches and diplexers makes it possible to couple both transponders to each of the antennas, provided the relevant antenna is suitable for transmission/reception at the chosen frequency band (see fig. 1.7). For the purpose of the Radio Science investigation, an Ultra-Stable Oscillator (USO) is connected to TRSP#2 in order to increase the frequency stability of the down-link signal (see par. 2.2 and App. D.1). As from November 2006, an anomaly on the S-Band transmitting chain is reducing the S-Band radiated power

---
[2]The transponders are identical, they are redounded.

Instrument Name	Scientific Objectives
APXS	α-p-X-Ray spectrometer
CIVA	Panoramic camera and IR microscope
CONSERT	Comet nucleus sounding
COSAC	Evolved gas analyzer: elemental and molecular composition
MUPUS	Multi purpose sensor for surface and sub-surface science
PTOLEMY	Evolved gas analyzer: isotopic composition
ROLIS	Descent camera
ROMAP	RoLand magnetometer (ROMAG) Plasma monitor(SPM)
SESAME	Comet acustic surface Sounding experiment (CASSE) Dust impact monitor (DIM) Permittivity probe (PP)
SD-2	Drill, sample, and distribution system

Table 1.2: *Payload of the Rosetta lander "Philae".* ([28]).

by an amount of $14\,dB$. This was a fatal failure for the radio science experiments "Bistatic-Radar" (see par. 4.4), and "Solar Corona" (see par. 2.3.2), which had to be canceled after that date.

The Rosetta spacecraft is provided with a high-gain steerable antenna of $2.2\,m$ dish diameter, a fixed $0.8\,m$ medium-gain antenna and two omni-directional low-gain antennas. The same conception and design of the VEX USO[3] is also exploited for the Rosetta communication subsystem (identical to the VEX communication subsystem, except that the USO drives both TRSP's, see fig. 1.7) which utilizes S-Band up-link for commands and S- and X-Band down-link for telemetry and science data ([28]). Output power is $20\,W$ at X-Band (delivered by a TWTA), and $5\,W$ at S-Band (solid state amplifier). The gain of the main antenna is $\sim 40\,dBi$ at X-Band and $28\,dBi$ at S-Band ([73]).

---

[3]The first developed and implemented USO was the USO on-board Rosetta. The VEX USO is manufactured on the base of the Rosetta USO design.

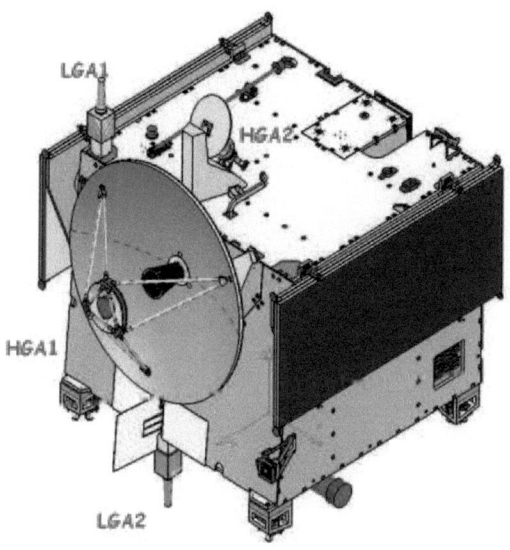

Figure 1.6: *S/C overview within reference frame. The four antennas are labeled* ([68]).

Communications are supported on ground by the ESA ground station in Cebreros (Spain), which consists of a $35\,m$ parabolic antenna with a gain of $68.2\,dBi$ in up-link and $66.5\,dBi$ in down-link (X- Band[4], [100]). The heart of the ground station communication system, which performs transmission and reception of the RF signals, is the IFMS (*Intermediate Frequency and Modem System*) ([82]). Besides the transmission of commands and the reception of telemetries, the ground equipment supports recording of Doppler- and range measurement[5], which are employed by the flight dynamics (FD) team in order to determine position and velocity of the spacecraft (tracking). (See App. A). Communications are indicated by the expression *TT&C*, which stands for *Tracking, Telemetry* and *Commands*.

---

[4]S-Band communications are not supported by the Cebreros G/S.
[5]A special *ranging* (RNG) tone is modulated on the carrier for ranging measurements.

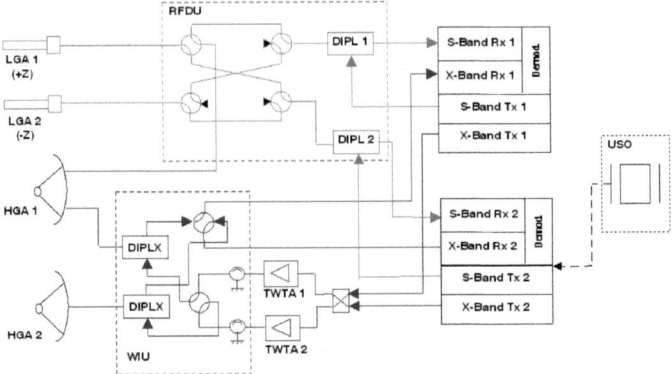

Figure 1.7: *Block diagram of the VEX S/C communication subsystem ([6]). The USO drives the TRSP#2 during the one-way Radio Science experiments (see par. 2.2 and App. D.1).*

The ESA Station in New Norcia (NNO, Australia), provided with a 35 m parabolic antenna with a gain of 66.0 dBi and 68.2 dBi at X-Band (up- and down-link, respectively) and 54.6 dBi and 55.8 dBi at S-Band (up- and down-link, respectively), ([100]), will serve as prime deep space ground station in all phases of the Rosetta mission. Furthermore, the NNO G/S supports the Radio Science experiment of both missions: VEX and Rosetta. For this purpose the VEX orbit was arranged in order to have visibility with the NNO station around pericenter, where the radio science observation are carried out because of the reduced distance of the spacecraft from the planet.

For some occultation observations the Radio Science experiments also rely on the ground stations of the *Deep Space Network* (DSN) of NASA. Some of these stations are 70 m in diameter with low system noise temperature ($\sim 25\,K$ and $\sim 20\,K$ at X- and S-Band, respectively).

1.3.1 Link-Budget

The radio link of a spacecraft mission must be carefully dimensioned by taking into account physical- and budget constraints. A proper design of the radio link must guarantee the required minimum value of the received signal power. This is determined by the sensitivity of the receiver and by the presence of thermal noise in the receiver electronics. It is common use to relate the power of the received signal, as measured at the input stage of the receiver, to the power of the thermal noise measured at the same point of the receiving chain in order to obtain a dimensionless quantity which is representative of the radio link quality. This quantity is given the name of *Signal-to-Noise Ratio* (*SNR* or *S/N*) and it is calculated as follows:

$$SNR = \frac{P_R}{N} \tag{1.1}$$

where P_R and N denote the power of the received signal and the power of the thermal noise, respectively[6]. Since information is conveyed by base-band signals modulating a high-frequency carrier, the spectral content of the received signal is comprised of the contributions of data and carrier. Considering the simple case of a single sinusoidal tone[7] modulating an RF carrier signal (FM or PM, see App. A), we shall define a link-budget for the carrier signal itself, indicated as S/N, and a link-budget for the sinusoidal tone, indicated as C/N[8]:

$$\frac{S}{N} = \frac{J_0^2(\beta) \cdot P_T \cdot G_T \cdot G_R}{L_{FS} \cdot N_0 \cdot B_c} \tag{1.2}$$

where J_n is the nth modified Bessel function of the first kind, β is the *modulation index*, P_T is the radiated power, G_T and G_R are the gain of the transmitting- and receiving antenna, respectively, L_{FS} is the *free-space* loss, N_0 is the noise power spectral density, and B_c is the bandwidth of the receiver set for reception of the carrier signal (see App. B).

---

[6]The term N also comprehends the back-ground noise, whose contribution is specified by the *equivalent antenna temperature* (see App. B).

[7]The chosen example might seems to have scarce practical relevance, since a single sinusoidal tone can convey a very small amount of information. Nevertheless, by considering that a signal can be thought of as made up of single sinusoidal tones (see Ch. 6), the developed theory can be applied to a vaster category of signals.

[8]A more comprehensive treatment of the topic is given in App. B.

The link-budget calculation for the data (i.e. single sinusoidal tone) is

$$\frac{C}{N} = \frac{2 J_1^2(\beta) \cdot P_T \cdot G_T \cdot G_R}{L_{FS} \cdot N_0 \cdot B_{data}} \qquad (1.3)$$

where B_{data} is the (two-sided) bandwidth of the receiver, set for proper reception of the data, in this case the transmitted sinusoidal tone.[9]

It must be noticed that additional losses, not included in (1.2), and (1.3), have also to be taken into account when dimensioning a radio link. These degradations can be due to various causes, such as mismatching in the transmission lines (i.e. transmitter-antenna or antenna-receiver), inaccuracy in the antenna pointing, and, generally, some characteristics of the circuitry which are not modeled by the system temperature.

A basic link-budget for the case of pure carrier signal for the ESA ground station Cebreros is presented in the table 1.4. The general data are given in tab. 1.3. The maximum distance Venus-Earth ($1.7 AU$) was considered for worst condition case. All calculation refer to the main high-gain antenna of the S/C, the *HGA1*. (See [6], and [100]).

	X-Band	
	G/S	S/C
D	$35\,m$	$1.3\,m$
T_{sys}	$51.3\,K$	$229.6\,K$
$T_{sys}\|_{dB}$	$17.1\,dBK$	$23.6\,dBK$

Table 1.3: *General data for VEX essential X-Band link-budget for the Cebreros ground station. (D is the diameter of the antenna dish, and T_{sys} is the system temperature.)*

---

[9]Choosing a value of β near to one lets the power of the information signal be mostly comprised in the first sideband, so that the receiver bandwidth is minimized. See also App. A

	Up-Link	Down-Link
P_T	73.0 dBm	48.4 dBm
G_T	66.5 dBi	37.0 dBi
G_R	35.0 dBi	68.2 dBi
L_{FS}	277.7 dB	279.1 dB
N_0	$-175.0\ dBm/Hz$	$-181.5\ dBm/Hz$
P_R	$-103.2\ dBm$	$-125.5\ dBm$
SNR_0	71.8 $dBHz$	56.0 $dBHz$

Table 1.4: *VEX essential X-Band link-budget for the Cebreros ground station ($f_{U/L} = 7.2 GHz$, $f_{D/L} = 8.4 GHz$). Distance: 1.7 AU.*

Chapter 2

The Venus Radio Science Experiment "VeRa" and the Rosetta Radio Science Investigation "RSI"

2.1 Fundamentals of Radio Science

The Radio Science method has established itself over the last decades as a unique remote sensing technique for the study of some physical features of celestial bodies, including the Earth. This technique allows scientists to investigate planetary atmospheres and ionospheres, gravitational fields, roughness and dielectric properties of planetary surfaces and solar coronal plasma.

The experiment is conducted by letting the investigated object be on the radio path between a transmitter and a receiver (typically a space probe and a receiving ground station on Earth, respectively). It is based on the principle that a radio frequency signal propagating through a physical medium will experience variations of some of its parameters because of the interaction between the electromagnetic waves and the medium itself. Information on the investigated object can therefore be obtained by evaluating time-dependent changes of parameters of the received radio signal. It is also possible to carry out a radio science observation when the object of the investigation does not directly lie on the radio path between transmitting and receiving antenna: this is the case for analyses of gravity fields, as the gravitational forces acting on the satellite alter the transmitted signal, thus allowing conclusions on the probed gravity field.

The Venus Express Radio Science experiment *VeRa*[1] is devoted to the study of:

- ionosphere: profiles of electron density vs. height above ground;
- atmosphere: profiles of temperature, pressure, and density as function of the height above ground;
- surface: dielectric properties and roughness (on a length-scale comparable with the signal wavelength);
- localized gravity anomalies;
- solar coronal plasma effects.

The almost polar orbit of Venus Express provides the opportunity to investigate the temperature structure of the atmosphere at all planetocentric latitudes, including day-night variations and signal absorption effects caused by gaseous components such as H_2SO_4, CO_2, and N_2. ([45], [46]).

Among the objectives of the Rosetta Radio Science Investigation *RSI* are ([73]):

- the determination of the cometary mass and bulk density, gravity coefficients, non-gravitational forces (gas-, and dust emissions);
- the determination of size and shape of the nucleus, its internal structure and the electromagnetic properties of the surface;
- the study of the abundance of mm-to-dm-sized dust grains near the nucleus, the inner coma plasma content, combined gas- and dust mass flux.

2.2 Instrumentation and operational modes

For the radio science investigation, the information about the inspected object is contained in the variations of some parameters of the received RF signal, such as

---

[1](**Venus Radio Science**)

frequency, amplitude, polarization, group delay. Such modifications,which are due to the interaction of the transmitted RF signal with the explored medium[2], allow investigators to draw conclusions about the nature of the target of the observation. Hence, it is evident that the most suitable probing signal to be used for radio science is a pure sinusoid of constant amplitude and constant frequency. For this reason, the *VeRa*- and *RSI* experiments exploit the RF carrier signal which is normally used for communications between spacecraft and ground station, thus relying on the spacecraft communication subsystem. In order to preserve spectral cleanliness, all possible modulating signals (TM, TC or RNG data) have to be switched off when a RS observation takes place.[3]

In a radio science experiment, the variations of the frequency of the received signal represent a primary font of information. For this reason, besides guaranteeing an adequately low system temperature of the receiver at the G/S,[4] it is very important to minimize the phase noise of the frequency source both at the G/S- and at S/C side. On ground, the purpose is achieved by implementation of an Hydrogen Maser which drives the local oscillator (LO). The frequency stability of the H_2-Maser is of the magnitude order of 10^{-15} ($\Delta f/f$, *Allan Deviation*, see App. C). On-board, the needed stability of the frequency source is achieved by implementation of an *Ultra Stable Oscillator* (USO, see App. D.1), whose Allan Deviation has proved to be better than 4×10^{-13} for integration times comprised between $10\,s$ and $100\,s$ for both, *VeRa* (see [43] and App. D.1), and RSI (see [73]). When a radio science observation is performed, the relevant transponder switches from its internal reference source, the TCXO (thermal controlled crystal oscillator), whose stability $\Delta f/f = 10^{-9}$ is too poor for science purposes (see Ch.8), to the USO.

Originally, the VEX USO should have driven both transponders. However, during the integration phase, some problems of electromagnetic compatibility have shown, which resulted in the difficulty for TRSP#1 to achieve lock when the USO was switched on. On the base of the troubleshooting activity, performed at the AS-

---

[2]In the case of the *gravity* experiment, the signature imposed by the mass of the investigated object on the signal frequency (Doppler effect) can be regarded as the effect of an *indirect* interaction between the transmitted RF signal and the medium.

[3]An exception to this rule is represented by the solar corona sounding experiment, when the observable is the group velocity (see par. 2.3.2). In that case the modulation of ranging tones is required.

[4]In each of the foreseen operational modes of the RS investigation, the employed RF signal is received and stored on ground, independently of which the transmitting source is (S/C or G/S). This concept will be clarified in the following.

TRIUM facility at Toulouse, it was decided to connect the USO only to TRSP#2, whereas the second output connector of the USO was terminated into a 50 Ω load (see fig. 1.7) ([44], [66]).

On ground, the functionalities of the communication system have been enhanced in order to support the radio science experiment. Besides the generation of the up-link- and the reception of the down-link signal, the IFMS #1, #2, and #3 provide *Doppler*, *Ranging*, AGC^5, and *Meteo* data, which constitute the so-called *Closed Loop* (CL) data. Furthermore, the IFMS equipment was augmented for reception and recording of *Open Loop* (OL) data, which complement the CL data for the radio science investigation. In particular, the IFMS#3 is dedicated to Radio Science in order to provide special settings of the receiver.

For many observations the radio science experiments also rely on the ground stations of NASA-JPL Deep Space Network (DSN). Some of these stations are equipped with 70 m dishes and the state-of-the-art receivers have a very low system noise temperature (25 K and 20 K at X- and S-Band, respectively, [68]).

Four different configurations of the radio-link, known as *transmission modes*, were specified by the radio-science team and made available by the ESA for the Radio Science investigation. They are (see fig. 2.1):

1. ONE-S (*One-Way Single*) in which the S/C sends a down-link signal at one frequency (S- or X-Band). In this mode the USO is the frequency reference source.

2. ONE-D (*One-Way Dual*). The same as ONES, with D/L signal sent on both frequencies.

3. TWO-S (*Two-Way Single*). This is a *coherent* communication mode, by which an up-link signal is sent from the G/S, received on-board, frequency converted, and transmitted back by the S/C. This mode is called coherent because the transponder acts as a mirror, conserving the phase information. In this case, the stability of the frequency reference source is governed by the G/S H_2-Maser (frequency stability $\Delta f/f \sim 10^{-15}$).

---

[5]**A**utomatic **G**ain **C**ontrol, a circuitry which monitors the incoming power and tunes the value of the amplification accordingly, in order to guarantee constant signal level to the receiver chain.

4. TWO-D (*Two-Way Dual*). The same as TWOS, with D/L signal sent on both frequencies.

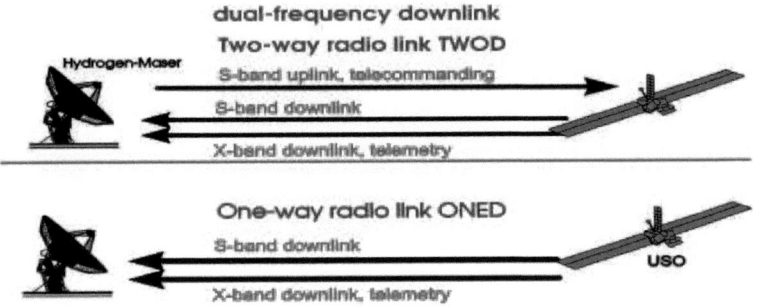

Figure 2.1: *Transponder configurations and transmission modes of the satellite. The Ultra-Stable Oscillator (USO) serves as on-board reference frequency source in the One-Way modes. In the Two-Way modes the reference frequency is generated by the G/S H_2-Maser (frequency stability $\Delta f/f \sim 10^{-15}$) ([45]).*

2.3 Brief description of the Radio Science Experiments

The basic idea of the radio science method –acquirement of information over remote targets by evaluating possible modification of some parameter of a radio signal– gives rise of a broad spectrum of possible applications, making radio science a very versatile remote sensing technique. A variety of different physical media can be investigated by means of radio science by exploiting different observation geometries.

In the frame of the Venus Express mission, the *VeRa* radio science investigation is comprised of four categories of observations, or experiments (see fig. 2.2). They are:

- the *occultation* experiment (OCC), in which the object of the investigation are the ionosphere and atmosphere of Venus (down to altitudes lower than

40 km above the surface). The occultation experiments, as the name suggests, exploits a geometry in which the satellite, as seen from the Earth, disappears behind the Venus at a certain point of its polar orbit and then re-emerges from the opposite side of the planet. The transmission mode is ONE-D.[6]

- the *bistatic radar* experiment (BSR), in which the object of the investigation are the dielectric properties of the Venus surface and the roughness at a scale comparable with wavelength of the employed radio signals. The observation geometry requires the S/C antenna beam to be directed toward the planet in such a way that the reflected signal can be collected at the Earth. Also for the BSR experiment, as for the OCC experiment, the transmission mode is ONE-D.[7]

- the *gravity* experiment (GRA), in which the RF carrier signal sent by the G/S is coherently sent back by the S/C (TWO-D). This experiments requires the radio-path between satellite and ground station to be free in order to evaluated the signatures impressed on the radio signal by gravitational force.[8]

- the *solar corona* experiment (SCO), in which the Sun must lie on the radio-path between satellite and ground station (within certain elongation angles), so that the RF signal can propagate through the solar corona. Object of the investigation is the the structure of solar coronal plasma. Transmission mode is TWO-D.[9]

The Rosetta radio science investigation (RSI) exploits the same four observation modalities, which allow the study of: ([73])

---

[6] The transmission of two different frequencies allows separation of dispersive refraction effects (caused by the ionosphere of the target planet) from the neutral atmosphere. After failure of the S-Band link (see par. 1.3), the occultation observations have being performed in One-Way single mode, exploiting the fact that at Venus the ionosphere and atmosphere are quite well separated, (see par. 5.2.2, [72], [94]).

[7] S-Band failure was fatal for the BSR experiment, as the strong atmospheric absorption of the X-Band signal does not grant levels of SNR which are suitable for the investigation.

[8] The effects of the S-Band failure on the GRA experiment depend on various parameters which have to be analyzed in order to draw conclusions.

[9] S-Band failure was fatal also for the SCO experiment, since the S-Band signal is more sensitive to the effects of charged particles.

- *cometary mass- and bulk density, gravity coefficients, moments of inertia, non-gravitational forces.* This suite of studies -based on the observation of perturbations of the S/C orbit by means of Doppler- and Ranging data- falls under the category of the *gravity* experiment;

- *size, shape, and internal structure of the nucleus.* These characteristics of the cometary nucleus are observed by means of *occultation* experiments. Comparing repeated occultation profiles, carried out at different relative positions between the S/C and the comet, the extension and the form of the occulting nucleus can be determined. Furthermore, it is expected that the nucleus, or at the least its external layers, will be penetrable by microwaves. This will give the possibility to study the bulk refractive index of the nucleus.

- *scattering properties of the nucleus*: carried out by means of a *bistatic radar* configuration;

- *coma investigations*. Scientific observation targets are:
    - the abundance of dust grains; hereby the observables are the signal attenuation (expressed as optical depth $\tau(\lambda)$) and the Doppler shift of the scattered signal. The optical depth at X-Band ($\lambda = 3.6\,cm$) reveals grains of the size of a few mm, whereas the differential optical depth, $\tau(3.6\,cm) - \tau(13\,cm)$, where $\lambda = 13\,cm$ is the wavelength of the S-Band signal, accounts for grains up to dm size. The experiment geometry is of the type of an occultation experiment, The signal impinging on the dust grains undergoes near-forward scattering, which implies a Doppler shift and frequency spreading of the scattered signal w.r.t. the direct signal (see paragraphs 5.1, and 4.3.2).
    - the plasma content;
    - combined gas- and dust mass flux: gas and dust impinging on the S/C surface impress radial accelerations which causes orbit perturbations detectable as Doppler shift of the received radio signal.

- *solar corona*; secondary science target. The experiment is conducted during the solar conjunctions happening in the cruise phase.

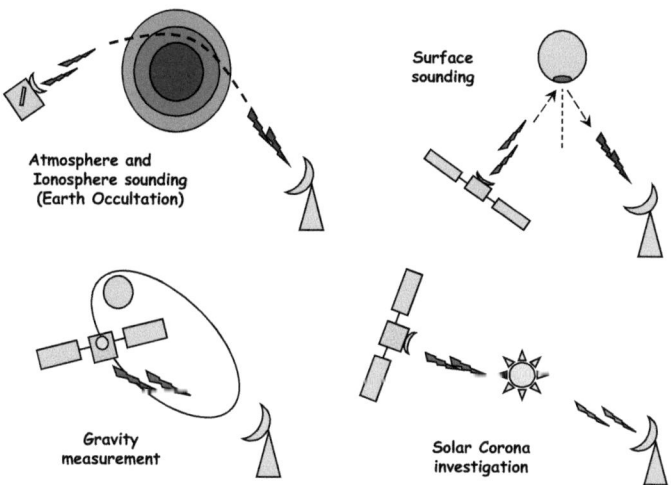

Figure 2.2: *The four observation geometries of the Radio Science Experiment* "VeRa".

2.3.1 The "Gravity" Experiment (GRA)

The objective of the *gravity* experiment (GRA) is the investigation of anomalies in the planet's gravity potential in particular regions. The experiment is based on the Doppler effect,[10] as the gravitational forces acting on the satellite continuously modify the relative velocity[11] of satellite and ground station. This results in time-dependent variations of the frequency of the received radio signal, which are evaluated in order to draw conclusions about the investigated gravitational anomalies.

To the purpose of evaluating gravity data, at first a model must be developed, which contains all known forces acting on the spacecraft. These are gravitational forces, due to the field interaction of the spacecraft with the mass of celestial bodies, and non-gravitational forces, i.e. the solar radiation pressure, due to the collision of the sun photons with the spacecraft (mainly with the surface of the solar arrays). From such a model is possible to calculated the expected frequency shift of the

---

[10]The Doppler effect is the change in frequency of a wave, which an observer in motion w.r.t. the source would perceive.

[11]The difference of the velocity vectors of S/C and G/S considered in the same inertial frame.

received radio signal, the so-called *predict*, which is then subtracted from the data. It is this difference, which is called frequency *residual*, that contains the looked-for information, that is, the unknown localized gravitational anomalies which are not comprised in the original force model.

The experiment is conducted in the two-way coherent mode (TWO-D), thereby relying on the ground station H_2-Maser as frequency reference source. The targets are located in the proximity of the orbital pericenter for experiment sensitivity reasons (minimum S/C altitude). The most important target of the *VeRa* GRA experiment is *"Atalanta Planitia"*, a low basin measuring $\sim 1500\,km$ in diameter and $2\,km$ in depth. Suitable orbits for GRA observations must be occultation-free and well separated in time from solar conjunction periods, in order for the gravitational signature exhibited by the frequency of the received signal not to be affected by spurious effects. The experimental method is sensitive to velocity changes as small as $\sim 10\mu m/s$ which can be interpreted as localized gravity effects and can contribute to the understanding of effects like mantle dynamics ([45], [46]).

2.3.2 The "Solar Corona" Experiment (SCO)

The solar corona is the outer, highly ionized part of the Sun's atmosphere, extending millions of kilometers into space. The coronal structure is highly variable over time and characterized by coronal holes, coronal streamers and rapid outburst events like Coronal Mass Ejections (CMEs). The Solar Corona experiment (SCO) is executed in order to investigate the structure of solar coronal plasma during the planetary superior conjunction phase (see fig. 5.1). The transmitted radio signal experiences a frequency shift during its journey through the coronal plasma. The magnitude of the frequency shift is related to a quantity called *"Total (Columnar) Electron Content"* (TEC), which represents the electron content of a column with a base surface of $1\,m^2$ and extending between transmitting and receiving antenna for the length of the optical path of the signal. Typically, the TEC is measured in *hexems* (1 hexem = 10^{16} electrons pro m^2, [46]).

As for the occultation experiment (see par. 5.2.1), the measurement is based on the different length of the optical path experienced by the signal w.r.t. the propagation in free space. Under the assumption of straight-line propagation in medium, the difference in optical path is expressed by ([40]):

$$\Delta l = \int_{l_0} (n-1)\, dl \; , \tag{2.1}$$

where l_0 is the signal path through free space. Variations of the travelled optical path are related to variations of the phase of the received signal:

$$\Delta \phi_m = \frac{2\pi}{\lambda_0} \int_{l_0} (n-1)\, dl \; , \tag{2.2}$$

which, in turns lead to a difference in frequency:

$$\Delta f_m = \frac{1}{2\pi} \frac{d(\Delta \phi_m)}{dt}. \tag{2.3}$$

For a ionized medium, the index of refraction is related to the signal angular frequency ω and to the plasma frequency ω_p by the relationship ([48]):

$$n(\omega)^2 = 1 - \frac{\omega_p^2}{\omega^2} \tag{2.4}$$

For $\omega \gg \omega_p$, which is for planetary and interplanetary plasma usually the case, eq. (2.4) can be expanded in series. By neglecting the contribution of higher order terms, the index of refraction can be expressed as: (dropping the explicit dependence on ω)

$$n = 1 - \frac{1}{2}\frac{\omega_p^2}{\omega^2} \; , \tag{2.5}$$

Taking into account that for the plasma frequency is:

$$\omega_p = \sqrt{\frac{N_e\, e^2}{m_e\, \varepsilon_0}} \; , \tag{2.6}$$

where N_e is the electron number density ($[1/m^3]$) of the medium, e is the electron charge, m_e is the electron mass, and ε_0 is the vacuum dielectric constant, eq. (2.5) can be rewritten as:

$$n = 1 - \frac{40.3 \cdot N_e}{f^2} \quad , \qquad (2.7)$$

where $f = \omega/2\pi$ is the signal frequency, and the numerical constant 40.3 has the dimension of $[m^3/s^2]$. Combing eqs. (2.3), (2.2), and (2.7) one obtains for the plasma induced frequency shift:

$$\Delta f_m = -\frac{40.3}{cf} \cdot \frac{d}{dt} \int_{S/C}^{G/S} N_e(l) \, dl \quad , \qquad (2.8)$$

where N_e depends on the optical path l, which extends between S/C and G/S, and c is the speed of light in vacuum. The integral $\int_{S/C}^{G/S} N_e(l)\, dl$ is the looked-for TEC, so that the observed Doppler frequency shift is proportional to the changes in the electron content of the observed region of the medium.

The total frequency shift experienced by the transmitted signal will also contain *non-dispersive* contribution due to the motion-related Doppler shift (classical Doppler effect). The difference Δf between the received frequency f and the transmitted carrier frequency f_0 will be therefore:

$$\Delta f = -\frac{f_0}{c}\frac{dl}{dt} + \frac{40.3}{c}\frac{1}{f_0}\frac{d}{dt}\int_{S/C}^{G/S} N_e(l)\, dl \quad , \qquad (2.9)$$

where dl/dt is the rate of change of the distance between spacecraft and receiver (relative velocity). The first term on the right-hand side of eq. (2.9) is the classical Doppler shift (linear in f_0 in first order). Making use of two coherently transmitted down-link frequencies allows one to separate non-dispersive effects from the plasma-induced frequency changes. In the case of VEX and Rosetta, the constant transponder ratio for the coherent dual frequency link is $880/240 = 11/3$. Indicating with subscripts S and X respectively S- and X-Band, the *differential Doppler* shift is defined as:

$$\Delta f_S - \frac{3}{11}\Delta f_X = -\frac{dl}{dt}\frac{f_S}{c} + \frac{40.3}{c}\frac{1}{f_0}\frac{d}{dt}\int_{S/C}^{G/S} N_e(l)\,dl + \frac{3}{11}\frac{dl}{dt}\frac{f_X}{c} +$$
$$-\frac{3}{11}\frac{40.3}{c}\frac{1}{f_X}\frac{d}{dt}\int_{S/C}^{G/S} N_e(l)\,dl =$$
$$= \frac{40.3}{c}f_S\left[\frac{1}{f_S^2} - \frac{1}{f_X^2}\right]\frac{d}{dt}\int_{S/C}^{G/S} N_e(l)\,dl \quad (2.10)$$

The dual frequency measurement thus provides the time derivative of the TEC and, by integration, the variations of the plasma density.

The *absolute value* of total electron content can be determind from the differential propagation delay (group delay) between both D/L frequencies (ranging signals) as ([73]):

$$\tau_S - \tau_X = \frac{40.3}{c}\left[\frac{1}{f_S^2} - \frac{1}{f_X^2}\right]\int_{S/C}^{G/S} N_e(l)\,dl\ , \quad (2.11)$$

where τ_S and τ_X are the propagation delay of the S-Band- and X-Band signal, respectively.

The two methods have different sensitivities: assuming a frequency uncertainty of $10\,mHz$ in eq. (2.10) (worst case conditions: distance: $1.7\,AU$, G/S: NNO, see [43]) leads to an uncertainty of 10^{-4} hexem, whereas an uncertainty of $1\,ns$ in the measurement of the group delay in eq. (2.11) implies an uncertainty of 10^{-1} hexem.

As mentioned in par. 1.3, the deterioration of the S-Band link caused solar corona measurements to be canceled, so that only the first SCO observation campaign could be successfully executed.

Chapter 3

The Radio Science Simulator (RSS)

In this chapter the Radio Science Simulator (RSS) will be briefly introduced. This is a software package which was conceived and developed at the Institute of Space Technology at the German Armed Force University in Neubiberg (Munich) as a basic computational tool for the analysis, planning, execution, and data evaluation of the Radio Science investigation ([39]). The data analysis software tools developed in the frame of this work are closely related to the RSS as either they exploit many of the computation functionalities of the RSS, or they need calculation from the RSS as input (for instance the prediction of the expected atmospheric frequency shift of the received signal during the mission), or even develop the RSS further by new functional blocks.

3.1 Concept & lay-out of the Radio Science Simulator

The RSS code is written in a MATLAB®/Simulink® environment and runs on the same platform. The central characteristic trait of this SW tool is its modular assembly, which allow maximum flexibility for enhancement and further development. Figures 3.1 and 3.2 show the structure of the RSS: the blue blocks represent independent pieces of SW (*modules*) which analyze different aspects of the experiment; they communicate by means of the *storage arrays* (yellow blocks), which serve as input/output units for the various modules. The coordination of the modules is provided by a *priority* factor assigned to each module. This guarantees the correct execution order of the modules.

3.1.1 Modules of the Radio Science Simulator

Hereafter the core modules of the simulator are described.

Time-Module

The first block of the RSS to be executed is the "Time-Module". As in a digital circuit, where each state transition of the elements happens on the edges of a clock signal, the simulation runs in step triggered by the MATLAB internal clock, which is capable of variable step size. The current value of the time step is added to the cumulative sum whose first element is the start time of the simulation, given as input in Julian Date (see after) format. The obtained time information is distributed to the other computation blocks.

Orbit-Module

The second block in order of execution is the "Orbit-Module". Position and velocity of the S/C around the target planet/celestial body are calculated at each simulation step also by taking into account the influence of the sun, planets and moons in the solar system and other effects, such as the solar wind. In order to increase the accuracy in all cases when not all intervening elements can be simulated, as for instance the spacecraft *wheel off-loading maneuver*, besides autonomous calculation, the orbit module is provided with the capability of reading and interpolate on a database providing the state vector of the S/C. This is the case for the Rosetta and Venus Express missions, where the simulations executed by the RSS are based on the orbit files provided by the ESA.

Ephemeris-Module

By means of the "Ephemeris-Module" is possible to obtain current position and velocity of defined celestial bodies of the solar system during the simulation. The calculation is based on the JPL DE 405 model of JPL (for a reference list see [86]). Effects of the finite propagation velocity of the light are accounted for, so that at each simulation step the Ephemeris-Module provides the RSS with the correct position vectors between S/C and G/S. This is achieved by means of a recursive calculation: first the propagation time for a given position of the bodies is computed; then the position of the bodies is updated to the elapsed time; after that the new propagation time is calculated; this leads to a new position and so on, until the algorithm converges within a specified error value.

Occultation-Module

This block calculates the interruptions of the *visibility* between S/C and G/S due to an interposed celestial body, typically the target planet. Thanks to the possibility of setting a variable step size for the simulation, the "Occultation-Module" can provide the exact time instant for the beginning (ingress) and the end (egress) of an occultation, this is, the time instants at which the line-of-sight between S/C and G/S strikes the outer limb of the atmosphere of the chosen planet (or the planetary disk), coming from the outer space or from behind the planet, respectively.

Ray-Correction-Module

Purpose of this module is the calculation of the required pointing of the S/C antenna in order to let the sent radio signal reach the G/S after propagation through the atmosphere of the interposed planet during an occultation event. It exploits the method of *ray tracing technique* applied on a model atmosphere derived by Jenkins from the Magellan Venus mission ([56], [62]). The data set implemented in the RSS assumes a spheric-symmetric atmosphere with increasing value of refraction index from outside toward the center, discretized in layer of $1\,m$ thickness. Due to the refraction in the atmosphere, the radio signal sent by the S/C is continuously bent in the atmosphere (see Ch.5). Adopting the perspective of the geometric optic, for each given position of the S/C along its orbit, the "Ray-Correction-Module" computes the bending angle of the relevant "ray" and the consequent antenna pointing ([79]).

Bistatic-Radar-Module

For each given position of the spacecraft- and ground station in the reference system, the "Bistatic-Radar-Module", or "BSR-Module" performs the computation of the coordinates of the "specular point"[1]. The set of all specular points which can be seen during the observation slot forms a "specular track" on ground. Considering this line as the subsequent positions of a single running point, it is possible to define a velocity of the specular point on ground, which has to be considered for the computation of the expected frequency shift in the bistatic radar experiment (see par.4.11).

Attitude-Control-Module

The "Attitude-Control-Module" transform the computed antenna pointing, provided either by the Occultation-Module or by the Bistatic-Radar-Module, depending on the experiment which has been simulated, in a time series of *quaternions*, which are written in a file and derived to the ESA for check and implementation on-board. In addition to that, the Attitude-Control-Module is capable of optimize the pointing in reference to given illumination constraints (see par.1.1.3), that is, the S/C is rotated around the antenna boresight in order to obtain the best illumination conditions while simultaneously keeping the required pointing (see also ([68])).

Doppler-Module

The "Doppler-Module" is in charge of the computation of the expected Doppler frequency shift due to the relative motion between the transmitting- and the receiving antenna. The computation accounts for relativistic effects, which can not be neglected at the implicated velocities ([47, 40]). (See also par. 4.11). This block also provides the RSS with the functionality of the prediction of the expected media-induced frequency shift, whose calculation is based on the pointing provided by the Ray-Correction-Module. Both, straight-line- and atmospheric predict are used as input for the Open Loop software (see 7.1.1)

3.2 Time Bases and Reference Systems

A clear understanding and a correct usage of the large variety of the conventional time- and reference systems is indispensable for each space mission ([40]). The RSS

---
[1]The "specular point" is the geometric point on the surface of the investigated celestial body which, under the assumption of "specular reflection" (the reflection angle equals the incidence angle) reflects the incoming wave in the direction of the G/S. (See pars.4.1, and 4.2)

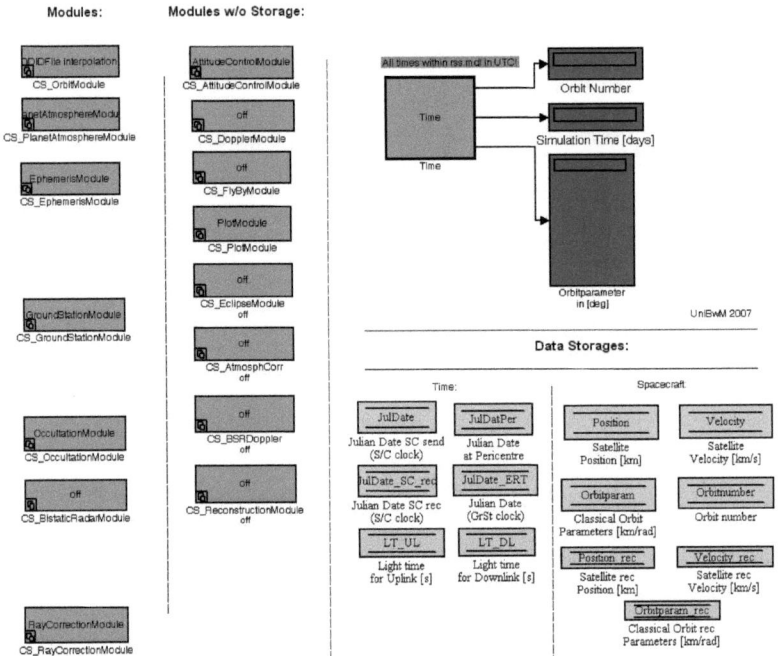

Figure 3.1: *Lay-out of the RSS (part a)*

can handle different frames and time bases and the relevant transformations, needed to convert quantities from one system to another. A short overview of some of them is given hereafter.

3.2.1 Time Bases

A *time-standard* defines a strategy for time measuring; it can define the time *unit* and the time *rate*, a specific *epoch*, or a time *scale*, thus specifying a method for measuring divisions of time. Among the conventional time standards which are commonly in use (see fig. 3.3):

- **Julian Date (JD)** The interval of time in days and fractions of a day since January 1, 4712 B.C., Greenwich noon, based on a given time scale (i.e., TT,

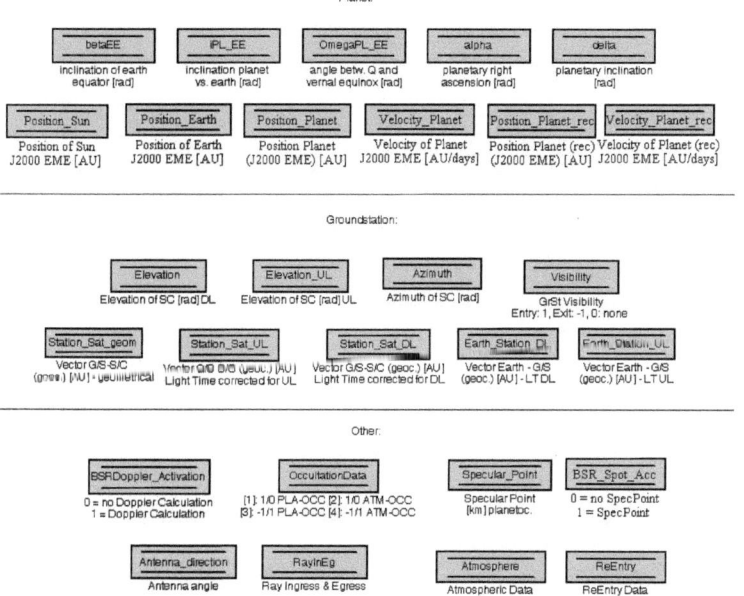

Figure 3.2: *Lay-out of the RSS (part b)*

see after).

- **Modified Julian Date (MJD)** Since the JD on January 1, 2000 had already grown to $2,451,545$ days, taking into account that the precision of usual computers is nowaday 15 decimal digits, the achievable accuracy in time is ms ([86]). In order to obtain better time resolution, it was retained sensible to reduce this number by subtracting the amount of days reached to a certain date, or *epoch*, i.e. November 17, 1858, at $00:00$, so that $MJD = JD2400000.5^d$. In the RSS a MJD is used, which refers to January 1, 2000, at 12:00. (See 8.2.6).

- **Ephemeris Time (ET)** A former standard astronomical time scale adopted in 1952 by the International Astronomical Union (IAU) and superseded in the 1970s. Ephemeris time was proposed in order to overcome the drawbacks of irregularly fluctuating mean solar time; it was a first application of the concept of a *dynamical time scale*, in which the time and time scale are defined implicitly, inferred from the observed position of an astronomical object via the dynamical theory of its motion.

- **Universal Time (UT1)** Medium solar time with constant average length of the day of 24 hours. Variations in the Earth revolution and rotation result in a not constant length of the second.

- **International Atomic Time (TAI)** Practical realization of a *uniform* time scale measured on the Earth surface by means of atomic clocks; the unit of TAI in the SI^2 is *seconds* ([s]).

- **Global Positioning System (GPS)** Reference time of the American satellite navigation system; introduced at the beginning of operations, runs parallel to TAI.

- **Terrestrial Time (TT)** Terrestial Time, formerly Terrestrial Dynamical Time (TDT). Uniform time scala with a constant difference to TAI. TT is the independent variable of geocentric ephemerides. TT replaced Ephemeris Time (ET) in 1984.

- **Universal Time Coordinated (UTC)** Time standard based on atomic clocks measurements, as for TAI, but corrected by *leap seconds*, in order to maintain the time difference to the UT1 system within $0.7\,s$. It is: $TAI = UTC + LS$, where LS is the sum of all introduced leap seconds.

- **Terrestrial Time (TCG)** Geocentric Coordinate Time TCG represents the time coordinate of a four dimensional reference system and differs from TT by a constant scale factor.

- **Barycentric Dynamic Time (TDB)** Relativistic coordinate time^3 scale applied to the solar-system-barycentric reference frame. It was first defined in

---

^{2}International System of Units, *Systme international d'units*.

^{3}In the theory of relativity the time is not an independent variable but the fourth coordinate of a four-dimensional reference system. The time specified by the time coordinate is referred to as *coordinate time* in order to distinguish it from *proper time*, which is time measured by a single clock between events that occur at the same place as the clock. Coordinate time depends not only on the events but also on the motion of the clock between the events.

1976 as a successor to the (non-relativistic) former standard of ephemeris time (ET).

- **Barycentric Coordinate Time (TCB)** Coordinate time standard used as independent variable for all calculations pertaining to orbits of planets, asteroids, comets, and interplanetary spacecraft in the Solar system. It is equivalent to the proper time experienced by a clock at rest in a coordinate frame co-moving with the barycenter of the Solar system: that is, a clock that performs exactly the same movements as the solar system but is outside the system's gravity well. It is therefore not influenced by the gravitational time dilation caused by the Sun and the rest of the system, as for the TDB time scale.

- **Dynamical Time Scale for the JPL DE 405 Ephemeris** (T_{eph}) During the years 1984 - 2003 the time scale of ephemerides referred to the barycenter of the solar system was TDB. Since 2004 this time scale for the JPL DE 405 ephemeris was be replaced by T_{eph}, which is approximately equal to TDB, but not exactly. On the other hand T_{eph} is mathematically and physically equivalent to TCB, differing from it by only an offset and a constant rate. In the RSS it is assumed: $T_{eph} \sim$ TDB, as this is within the required accuracy .

Figure 3.3 shows the difference in seconds among the mentioned time-standards.

3.2.2 Reference Systems

The position of an object in space is given through its *coordinates*, which express the distance of the object from the origin of a specified *reference system*.

As main inertial reference system, the RSS makes use of the *International Celestial Reference System* (ICRS), the current standard celestial reference system adopted by the IAU. This is a frame centered in the barycenter of the solar system, whose axes are intended to be "fixed" with respect to space. The $\vec{X}$ axis is directed toward the *vernal point*; the plane it forms together with the $\vec{Y}$ axis is the same plane of the Earth equator in the year 2000[4]. Therefore this reference system is called "BC J2000 EME": BC stands for barycentric, and J2000 EME stands for *Earth Mean Equator* on January 1, 2000.

---
[4]Due to axial precession and nutation the position of the equatorial plane w.r.t. the ecliptic plane changes over time.

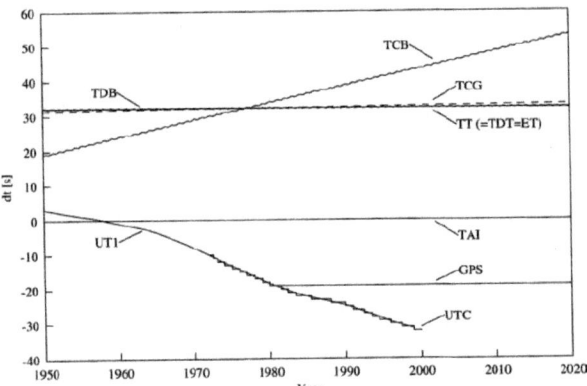

Figure 3.3: *Difference in seconds among different time-standards from* 1950 *to* 2020 ([85]).

The position of some objects of interest could be specified in a "local" frame, rather than in the BC J2000 EME, as, for instance, the position of the feed of the High-Gain-Antenna of the spacecraft, which is known in the reference frame centered in the S/C. Another example is the position of the ground station antenna feed on the Earth, or the location of the *specular point* (see 4.1) on the surface of the target planet, or some points of interest in an occultation entry or exit (see 5.1). All this points are specified in not inertial frames, as, for instance, a *planetocentric* reference system. Such a frame has its origin in the center of a considered planet and proper orientation, specific for each planet. It is fixed with the planet, so that objects on the surface on the planet don't change their coordinates due to the motion of the planet in space. An example of planetocentric reference system is the *International Terrestrial Reference System* (ITRS).

In order to perform vector summation, all vectors of a system must be expressed in the same frame. The RSS is provided with opportune libraries for the conversion.

3.2.3 Reference frame and time basis of input data

Of central importance when processing the data is the homogeneity of reference frames and time bases in the whole processing chain.

Along with the raw data, whose time stamps are specified in UTC ([55]), the Open Loop data processing software assumes as input the straight-line- and atmospheric frequency predicts delivered by the RSS. These are, in turns, based on the ESA orbit files, as mentioned in par. 3.1.1, where the S/C state vector is specified in the J2000 inertial reference frame, with Barycentric Dynamical Time (TDB) as independent variable ([20]). The RSS converts all times in UTC format for the computation, but delivers the frequency predicts with time in the TDB format. Therefore, after input acquisition, a conversion from TDB to UTC takes place in the OL SW, which, analogously to the RSS, operates in UTC.

Chapter 4

The Bistatic Radar Experiment

4.1 The concept of the "Bistatic Radar" experiment

The *bistatic radar* experiment (BSR) is devoted to the study of surface characteristics in regions of special interest. By aiming the main high-gain antenna of the spacecraft toward the "specular point" on the target body, it is possible to receive and measure the reflected power at the ground station as the spacecraft moves along its orbit. Estimates of the dielectric constant can be derived from the evaluation of the power received in two orthogonal polarizations and comparison with Fresnel reflection coefficients. From the dielectric constant (and suitable modeling) it is then possible to estimate the density of the surface material. Furthermore, frequency dispersion of the echo signal provides information about the roughness of the surface, whose *rms* slope can be determined at horizontal scales from a few tens- to some hundreds of wavelengths ([87]).

Figure 4.1 illustrates the Deep Space Network (DSN) ground station configuration for bistatic radar experiments: four independent channels allow reception, sampling, and recording of echo signal in both polarization senses for S- and X-Band.

It is possible to implement the bistatic radar geometry with at least three different S/C attitudes, which are called "pointings". These are the *specular-pointing*, the *spot-pointing*, and the *inertial-pointing* ([86]).

In the specular-pointing mode, at each given time t_0 the S/C antenna bore-

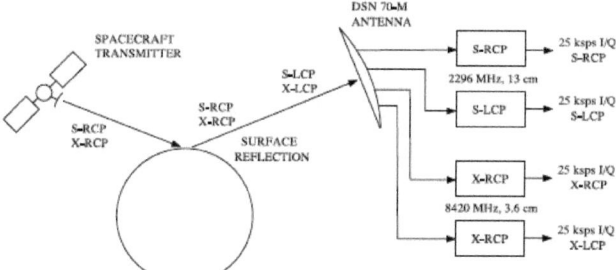

Figure 4.1: *DSN ground station configuration for bistatic radar experiments; by the implementation of four independent receivers the system is capable of coherently sampling and recording signal at S- and X-Band at both, right- and left circular polarization (RCP and LCP, respectively).* ([88]).

sight is directed toward the specular point on ground, i.e. the point which gives specular reflection in the direction of Earth. As the relative position of the involved bodies (target planet, S/C, and G/S) changes during the experiment, the S/C is continuously steered while flying over the region of interest, in order to let the antenna beam follow the specular point on ground. The specular points line up in a track on the surface of the planet, which is called "specular track".

Instead of following a line of specular points on ground, in the spot-pointing configuration the antenna beam is kept on the observation target on the surface for the whole duration of the pass, while the back-scattered echo signal is collected at the Earth.

The inertial-pointing is used when no slew maneuvers can be executed during the observation. In this case the antenna is pointed toward the planet surface at the beginning of the experiment and the S/C remains in the same attitude for the whole duration of the pass. The initial pointing direction must guarantee that the target point is contained in the track formed by the projection of the antenna beam on ground and that it will be met by the radio signal at the proper incidence angle, i.e. the angle which causes the reflected signal to propagate along the direction of the Earth.

In the frame of the *VeRa* investigation, bistatic radar observations are conducted in specular-pointing mode from positions in the orbit which are located in the vicinity of the pericenter, as the distance between S/C and planet is minimal. The proper S/C attitude, which shall take into account the refraction of the radio signal in the Venus atmosphere, is computed by the *VeRa* team by means of the RSS (see par.3.1.1). ([68], [45], [46]).

The same observation configuration is implemented in the Rosetta mission for the study of the surface properties of the nucleus at comet 67P/Churyumov-Gerasimenko. In addition, evaluation of some signal spectral features, such as Doppler shift and spectral broadening, will contribute to the investigation of cometary dust (see par. 2.3).

The case of BSR at Churyumov-Gerasimenko requires particular attention, as the small angular moment of the comet imposes particular conditions on the received signal. Details are given in the paragraph dealing with the algorithm for calculation of the expected Doppler frequency shift for bistatic radar observations (see par.4.5.1).

4.2 Experiment geometry

In a down-link[1] bistatic radar configuration, the bore-sight of the spacecraft antenna is pointed toward the observation target at the planet surface and transmits an opportune RF signal. Proper antenna pointing makes it possible to receive the reflected echo signal at the Earth ground station.

Fig. 4.2 show the bistatic radar geometry in the plain formed by the probe, the center of mass of the target planet, and the receiver. The position of a general scattering point at the planet surface in the planetocentric coordinate system is given by the vector $\vec{s}$, whereas vectors $\vec{t'}$ and $\vec{r'}$ respectively denote the positions of transmitter and receiver in the same frame.

Fig. 4.3 shows the general scattering geometry in the case of electromagnetic radiation impinging on a locally homogeneous and isotropic surface. In the ideal case of a perfect smooth surface, *specular reflection* would occur. In this case the

---

[1]Bistatic configurations for radar observations offer the possibility of two different implementations, namely *up-link* and *down-link*, depending on which of the two involved antennas is sending or receiving. ([87])

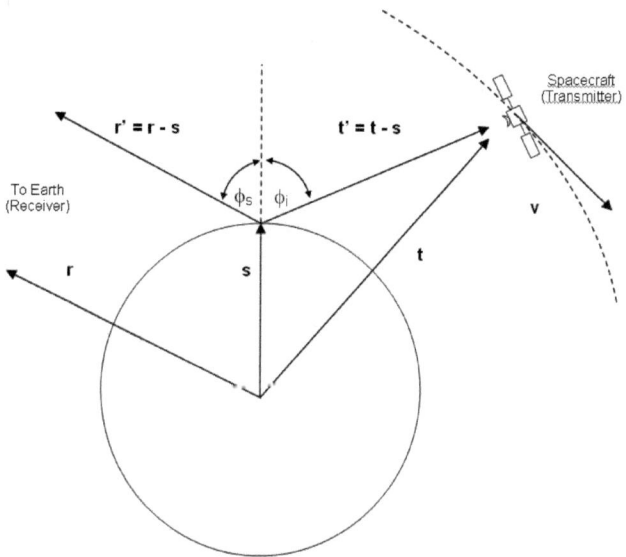

Figure 4.2: *Two-dimensional geometry for a bistatic radar observation (down-link configuration). In the planetocentric reference system, positions of transmitter (S/C) and receiver(G/S) are given by vectors $\vec{t}$ and $\vec{r}$, respectively. The S/C flies with velocity $\vec{v}$. Transmitter- and receiver positions w. r. t. the general scattering point, determined by the vector $\vec{s}$, are given by $\vec{t'} = |\vec{t}-\vec{s}|$, and $\vec{r'} = |\vec{r}-\vec{s}|$, respectively. The incidence angle is ϕ_i; the scattering angle is ϕ_s. The azimuthal angle ϕ_a between the incidence plane and the observation plane (see text for definition) is not shown. In the case of specular reflection, the two planes are the same ($\phi_a = 0$) and scattering (reflection) angle equals the incidence angle ($\phi_s = \phi_i$).* ([87])

incidence plane and the observation plane[2] coincide, that is, $\phi_a = 0$. The scattering angle is called *reflection* angle and equals the incidence angle, that is, $\phi_s = \phi_i$ (see App. E).

---

[2]Respectively, the plane containing the wave vector of the incident radiation and the vector normal to the surface of incidence and the plane containing the wave vector of the scattered energy and the vector normal to the surface of incidence.

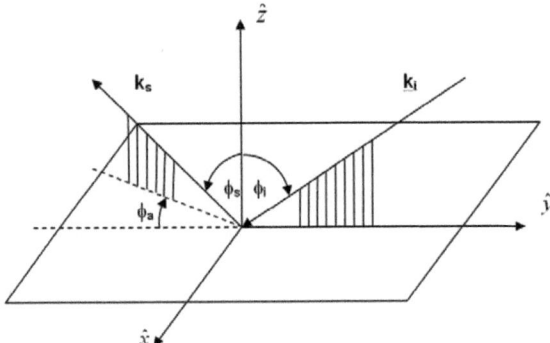

Figure 4.3: *General scattering geometry for locally homogeneous and isotropic surface.* $\vec{k}_i$ *and* $\vec{k}_s$ *are the wave vectors of the incident and scattered wave, respectively.* ϕ_a *defines the angle between the incidence plane and the observation plane (see text for definition). The conditions* $\phi_a = 0$ *and* $\phi_i = \phi_s$ *define "specular" scattering (specular reflection).* ([87])

4.3 Power frequency spectrum of echo signal

In order to infer surface properties from the echo signal, a physical optics model is put at the base of the VEX BSR investigation. This kind of models assume that the reflecting surface is homogeneous and isotropic, that it is gently undulating on dimensions equal to or larger than the radio wavelength λ of the transmitted RF carrier signal, that subsurface scattering is negligible, and that the surface radius of curvature is large with respect to λ ("tangent plane" approximation), ([87]). In this way, the experiment can be regarded as "quasi-specular", that is, surface elements in the vicinity of the specular point are assumed to have *tilt* angles and orientations such as to reflect the incident radiation toward the Earth. Therefore, the echo signal is comprised of the coherent summation of the fields reflected by the single scatterers, since the phases of the component fields are narrowly distributed around the phase of the echo from the specular point.

Fig.4.5 shows the frequency spectrum of a time average of the signal received during a VEX BSR experiment on 15th June, 2006. The spectrum of the received echo signal shows three typical characteristics:

- *Doppler-shift* or frequency drift;

- *dispersion*, or spectral broadening;

- *echo strength*.

4.3.1 Doppler-shift

The Doppler-shift of the echo signal spectrum is a time-dependent quantity. It depends on the continuous variation of the length of the radio path which goes from the spacecraft to the specular point and from the latter to the ground station as the satellite proceeds along its orbit. Referring to the geometry described in fig. 4.2, the frequency shift of the received echo signal can be expressed as ([87]):

$$f_s(t) = -\frac{1}{\lambda}\frac{d}{dt}\{|\vec{r'}(t)| + |\vec{t'}(t)|\} \tag{4.1}$$

It must be considered that, besides the reflected signal, it is very common during a BSR experiment to receive signal directly radiated by a side-lobe of the spacecraft antenna in the direction of the ground station. Similarly to the echo signal, also the direct signal will show a time-dependent Doppler shift with respect to the nominal frequency of the transmitted carrier frequency. Referring to the geometry of fig. 4.2, the Doppler frequency shift of the direct signal is given by ([87]):

$$f_d(t) = -\frac{1}{\lambda}\frac{d}{dt}\{|\vec{r}(t) - \vec{t}(t)|\} \tag{4.2}$$

By looking at eqs. (4.1) and (4.2), it can be seen that spectral contributions from echo signal and direct signal are separated and both drifting apart with respect to a fixed reference.

4.3.2 Spectral broadening

Frequency dispersion of the echo signal is due to the fact that the reflected energy is made up of the sum of single reflections arising from many elementary scatterers surrounding the theoretical specular point in the footprint of the antenna beam.

Any surface element contributing to the echo involves a different radio path (the geometrical path from S/C to reflector and from reflector to Earth, neglecting atmospheric/ionospheric bending effects), whose change rate slightly differs from the other ones. This causes each echo contribution to exhibit a different Doppler-shift, which results in a spectral broadening of the total echo signal. In the case of a perfectly conducting planetary surface with Gaussian roughness statistics, uniformly illuminated by both antennas, the width of the bell-shaped spectrum of the echo is directly proportional to the *rms* surface slope ξ: ([87])

$$B = 4\sqrt{(ln2)} \cdot \frac{V\xi}{\lambda} \cdot cos\varphi \tag{4.3}$$

where B is the half-power width of the echo, V is the velocity of the specular point on ground, φ is the incidence angle, and ξ is the adirectional *rms* surface slope in radians. ξ is a very important parameter for characterizing planetary surfaces.

4.3.3 Echo strength

The amount of signal power received at the G/S from the reflection is given in the general case by (dropping the time dependence): ([87])

$$P_R = \int_S \frac{P_T G_T(\vec{t}, \vec{s})}{4\pi |\vec{t^2}|} \cdot \sigma_0(\vec{r}, \vec{s}, \vec{t}) \cdot \frac{\lambda^2 G_R(\vec{r}, \vec{s})}{(4\pi)^2 |\vec{r^2}|} \, dS \tag{4.4}$$

where P_T is the transmitted power, G_T is the gain of spacecraft antenna in the direction defined by $\vec{t}$ and $\vec{s}$, σ_0 is the *specific radar cross-section* of the illuminated surface S, G_R is the gain of the receiving antenna in the direction defined by $\vec{r}$ and $\vec{s}$. The integration is performed over the surface S which is mutually visible from both transmitter and receiver.

In practice, there is an "effective" scattering area A_{eff} which is responsible for most of the quasi-specular echo. If the transmitting- and receiving antennas do not illuminate the whole surface, but cover the effective area, then eq. (4.3) is still a good approximation. In the case of backscattering (mono-static radar), the effective area is a spherical cap of radius r_{eff} within which the elementary scatterer are conveniently oriented for specular reflection. In fig. 4.4, the angle ϕ_i is the tilt angle required to the ith facet in order to give maximum contribution to the backscattering. The effective area is then given by: ([87])

$$A_{eff} = \pi \, r_{eff}^2 \qquad (4.5)$$

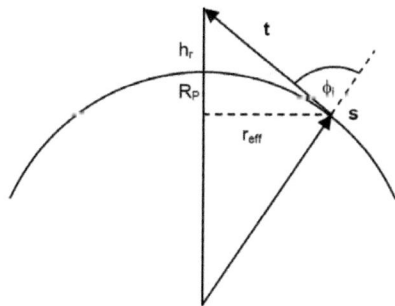

Figure 4.4: *Two-dimensional geometry for estimation of "effective" scattering area for mono-static radar. R_P is the planet radius, h_r is the radar distance from the surface, $\vec{s}$ is the vector from center of the planet to the ith elementary scatterer, $\vec{t}$ is the vector from the ith elementary scatterer to the radar, ϕ_i is the tilt angle required to the ith facet in order to give maximum contribution to the backscattering, and, finally, r_{eff} is the radius of the considered effective area. ([87]).*

Referring to fig. 4.4, for modest rms slopes ($\xi \sim 0.1$ radian) is: ([87])

$$r_{eff} \sim \begin{cases} 2 R_P \xi & h_r \gg R_P \\ 2 h_r \xi & h_r \ll R_P \end{cases} \qquad (4.6)$$

In the case of bistatic radar, the oblique incidence deforms the projection of the beam onto the surface into an ellipse of radial and transverse half axes a_r, and a_t, respectively, and approximate area:

$$A_i \simeq \pi \, a_r \, a_t \qquad (4.7)$$

In this case, the reduction in the received power can be estimated from the ratio of eq. (4.7) to eq. (4.5).

The received power is therefore: ([87])

$$P_{Robs} \simeq \begin{cases} P_R \cdot \dfrac{A_i}{A_{eff}} & A_i \leq A_{eff} \\ P_R & otherwise \end{cases} \qquad (4.8)$$

Typically, the intensity of the received echoes ranges from tenths of *zeptowatts* (zW) to hundreds of zW, where $1\,zW = 10^{-21}\,W$ ([90]).

4.4 Estimate of the dielectric constant of the planetary surface

The estimation of the dielectric constant from radar returns is instrumental in the characterization of the material(s) forming the soils object of the investigation. Since the dielectric properties of many kinds of natural materials are known, it is possible to infer about the composition of the probed terrains from the estimate of the dielectric constant. The scattering process, however, results from a variety of concomitant factors which have to be understood and modeled in order to draw reliable conclusions. First of all, depending on materials, density, mixing ratio, humidity, and, to a minor extent, temperature and frequency, have influence on the value of the complex dielectric constant ([11]). In addition to this, the *morphology* of the illuminated terrains plays an important role in the shaping of the echo power spectrum. By considering that the characterization of a surface as "smooth" or "rough", or the determination whether the reflection is ruled by "surface scattering" or "volume scattering" are relative to the adopted wavelength, it is evident that caution has to be adopted when inferring physical properties of the observed areas from spectral features of the radar echoes.

In the following, the case of Gaussian scattering law (scattering from a deep, gently undulating surface with Gaussian statistics and autocorrelation function)

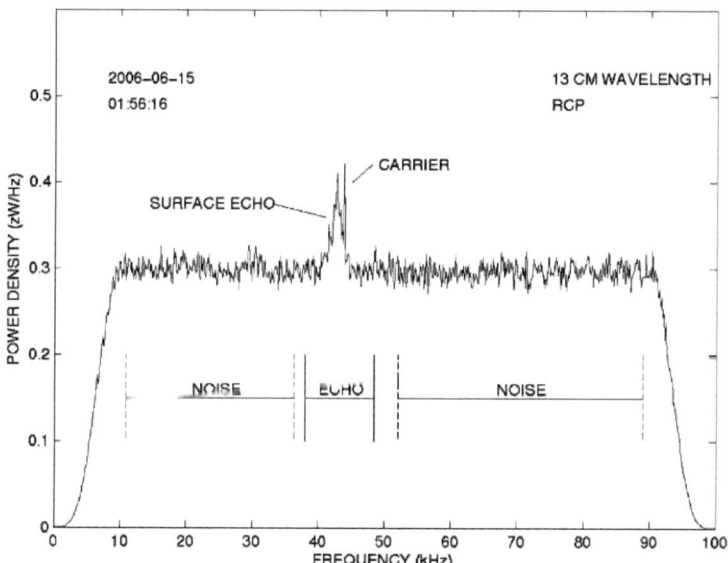

Figure 4.5: *Example S-Band power density spectrum from the VEX-VeRa BSR observation of June 15, 2006 (right circular polarization). The* 1024*-points* 100 kHz *spectrum is the average of 976 individual power spectra collected within an interval of* 10s *centered at time 01:56:16, when the specular point was in a plains area northwest of Maxwell Montes. The surface echo, identified by the horizontal bar over* 38-47 kHz, *is spread over a bandwidth of about* 10kHz. *Within the echo window a narrow carrier signal propagating directly from the spacecraft is visible at* 43 kHz. *In order to compute the echo strength, the baseline noise power density* N_0 *is calculated from the spectral regions on either side of the echo (identified by horizontal bars over* 10-37 kHz *and* 53-88 kHz, *respectively). The signal was acquired by DSN (see fig. 4.1).* ([90]).

will be assumed. This allows reflectivity (and, therefore, dielectric constant[3]) and roughness to be determined independently to a first order ([87]), since the total received power is only weakly dependent on slopes ([97]).

---

[3] Reflectivity relates the received power to the dielectric constant through the *Fresnel* reflection coefficients (see eq. (E.7)).

4.4.1 Reflection coefficients

When specular reflection occurs at a plane boundary between physical media of different refraction index, a fraction of the power of the incident signal is transmitted through the interface with a different propagation direction (refraction) and the remnant power is reflected back in the half-space of provenience under the same angle of the incident wave. Fresnel's coefficient relate the amount of transmitted and reflected power to the incidence angle and to the dielectric and magnetic properties of the involved media (see App. E).

In the special case of a planetary bistatic-radar experiment, at first neglecting planetary atmospheric/ionospheric contributions, one of the involved media is vacuum, whose relative dielectric and magnetic constants are unitary. Furthermore, except for special cases of terrains with magnetic features, the relative magnetic permeability of the planet crust can be considered to be unitary, as well. Under these assumptions, the reflection coefficients specified in eqs. (E.3), and (E.5) can be written as ([87]):

$$R_H = \frac{\cos(\phi) - \sqrt{\epsilon - \sin^2(\phi)}}{\cos(\phi) + \sqrt{\epsilon - \sin^2(\phi)}} \tag{4.9}$$

$$R_V = \frac{\epsilon \cdot \cos(\phi) - \sqrt{\epsilon - \sin^2(\phi)}}{\epsilon \cdot \cos(\phi) + \sqrt{\epsilon - \sin^2(\phi)}} \tag{4.10}$$

where the subscript "⊥" was substituted by the subscript "H" for horizontally polarized waves and the subscript "∥" was substituted by the subscript "V" for vertically polarized waves. When ϵ is complex (conductivity of the involved materials > 0), R_H and R_V are complex quantities. In case of circularly polarized wave investing the observed surface with incidence angle $\theta_i \neq 0, \theta_B$, where θ_B is the Brewster angle (see App.E), the reflected wave will exhibit circularly polarized components in both senses ([87]):

$$R_{SC} = \frac{R_V + R_H}{2} \tag{4.11}$$

$$R_{OC} = \frac{R_V - R_H}{2} \qquad (4.12)$$

where "SC" stands for *same circulation sense* and "OC" stands for *opposite circulation sense*. Therefore, when specular reflection occurs, the reflected power is proportional to the incident power through the Fresnel's coefficients (squared). Assuming right circular polarization for the radiated signal, the power of the reflected signal in both polarization senses is given by:

$$P_{RCP} = |R_{SC}|^2 \cdot P_i \qquad (4.13)$$

$$P_{LCP} = |R_{OC}|^2 \cdot P_i \qquad (4.14)$$

where *RCP* and *LCP* stand for *Right-Circular-Polarization* and *Left-Circular-Polarization*, respectively, and P_i indicates the incident power at the reflecting surface. Consequently, computing the ratio of the power received in both polarizations allows direct calculation of the dielectric constant, as the observation geometry is known. Figure 4.6 shows the dependence of ϵ on the incidence angle ϕ for different values of the ratio of the received power, whereas figure 4.1 illustrates the Deep Space Network (DSN) ground station configuration for bistatic radar experiments: four independent channels allow to receive, sample and record signals in both polarization senses for S- and X-Band.

As an example, figure 4.7 shows spectrograms of received S-Band signals from the VEX-VeRa bistatic radar observation at Maxwell Montes on June 15, 2006. In order to keep the signal in the displayed bandwidth a tuning offset of $29kHz$ was introduced at 01:57, causing the discontinuity in the signal tracks. The left panels represent time sequence of power density spectra of RCP- and LCP received signals (top, and bottom, respectively). The right panels show the time sequence of magnitude and phase of the cross-spectrum (top, and bottom, respectively), which is the product of the Fourier Transform of the RCP signal times the conjugate of the Fourier Transform of the LCP signal.

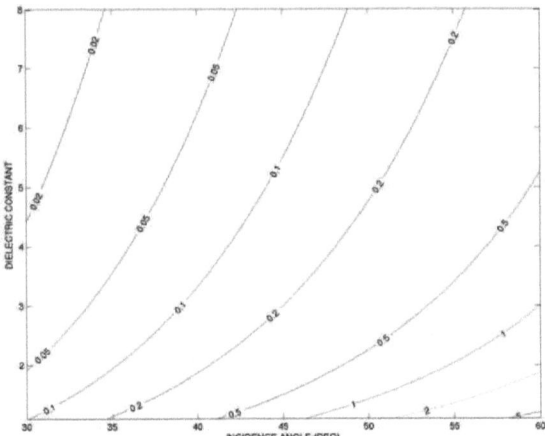

Figure 4.6: *Contours of RCP/LCP power ratio vs. incidence angle ϕ and dielectric constant ϵ,* ([89]).

The phase of cross-spectrum is of particular importance as it is sensitive to complex values of the dielectric constant, so that it is possible to asses whether the inspected region contains conducting materials or not. In the lower right panel, emerging from the background noise, the clear signature of the phase spectrum attests the coherency of the RCP and LCP echoes, thus confirming the validity of the quasi-specular assumption. Time variations of the phase difference between the RCP- and the LCP echo signal are due to the spacecraft rotation around the antenna bore-sight[4], plasma effects, and differences in the electronics of the two receiving chains ([90]). Such a slow phase drift would anyway not prevent the recognition of abrupt phase changes which would occur in the transition from non-conductive to conductive terrains and vice versa. Spectral broadening of the echo signal is also evident, particularly when compared with the sharp signature of direct RCP carrier signal, radiated by a side-lobe of the spacecraft high-gain antenna (upper-left panel). Unfortunately, RCP echo signal drops consistently at about 02:00 in correspondence of Maxwell Montes, compromising the ability to conclude about the dielectric constant in the region of interest. An anomaly in the S-Band transmitting

---

[4]The S/C rotation is imposed by the thermal protection policy. The maneuvers do not affect the pointing, as the spacecraft rolls around the antenna bore-sight.

chain on the S/C side (see par. 1.3) prevented further BSR observations, since, due atmospheric absorption in the Venus atmosphere, the X-Band echo results too weak for definitive conclusions to be drawn.

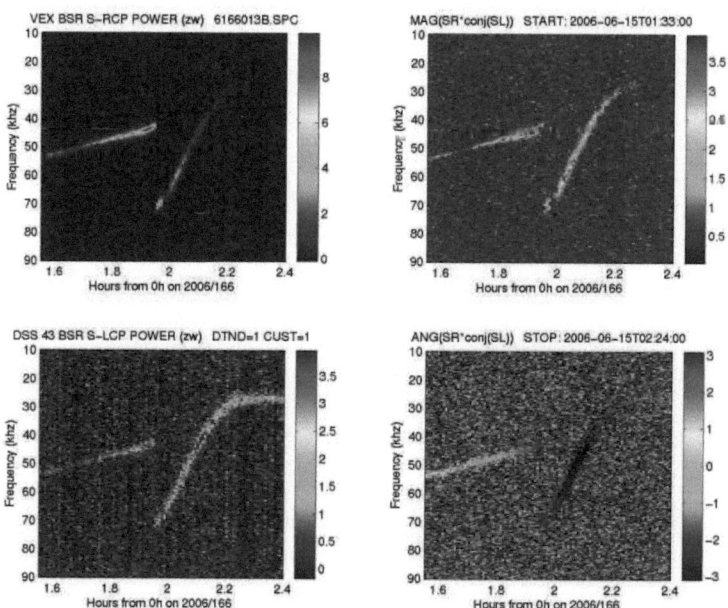

Figure 4.7: *Spectrograms of S-Band echo signals received during the VEX-VeRa bistatic radar observation at Maxwell Montes on June, 15th 2006. Upper-left and lower-left: RCP- and LCP time sequence of power spectra, respectively. Upper-right and lower-right: time sequence of magnitude and phase of the cross-spectrum, respectively. Color bars represent signal power in units of zeptowatt (zw) per 97.7 Hz frequency bin (1 zw = 10^{-21} W) except for the lower-right panel, where the side color bar represents phase values in radians. (Signal acquired by DSN).* ([89]).

4.5 Computation of expected Doppler frequency shift of echo signal

Besides accurate determination of the position of the specular point during the flight and precise calculation of the relevant spacecraft attitude needed to follow the specular track on ground (see par. 4.1), the successfully execution of a bistatic radar experiment also depends on proper settings of the receiving systems at the ground station. In particular, the determination of the receiver bandwidth is of central importance for the correct reception and sampling of the echo signal. To this purpose, the expected bandwidth occupation of the received surface echo must be evaluated. Among the spectral characteristics of the reflected signal described in paragraph 4.3, those which are relevant to this analysis are the Doppler frequency shift- and shift rate of the received echo signal.[5] In order to perform this calculation, a new module of the RSS was developed, the *"Bistatic Radar Doppler Module"*.

The frequency shift of the reflected- and direct signal is due, as foreseen by the Doppler effect, to the time changes of the distance traveled by the signal during the propagation, as in (4.1) and (4.2). Nevertheless, when high velocities are implied, as in space applications, relativistic effects can not be neglected ([47]). For this reason a more comprehensive formula was adopted, which takes into account also the effects of the General Relativity. Considering a signal transmitted by a platform S at the frequency f_S, a receiving platform E will measure a frequency f_E for the same signal because of the Doppler shift[6]. Considering the geometry of fig. 4.8, the received frequency can be expressed as a function of the transmitted frequency ([40]):

$$\frac{f_E}{f_S} = \frac{1 - \hat{n} \cdot \vec{\beta}_E + \frac{1}{2}|\vec{\beta}_E|^2 - \Phi_E/c^2}{1 - \hat{n} \cdot \vec{\beta}_S + \frac{1}{2}|\vec{\beta}_S|^2 - \Phi_S/c^2} \qquad (4.15)$$

where subscripts S and E denote the transmitting and the receiving platform, respectively. The direction from S to E is given by the unit vector $\hat{n}$, which is oriented in the direction of the signal propagation; $\vec{\beta} = \vec{v}/c$, where $\vec{v}$ is the coordi-

---

[5]Spectral broadening of the echo signal contributes only for a small percentage of the total bandwidth occupation.
[6]Coordinates and vectors are given in the barycentric inertial frame (see 3.2).

nate velocity of the platform and c is the speed of light in vacuum. The last terms in numerator and denominator of the right side of (4.15) account for the effects of General Relativity: Φ is the gravitational potential of the relevant platform (see also: [86]). In general, the gravitational potential Φ is given by:

$$\Phi = -\frac{GM}{r} \tag{4.16}$$

where G is the gravitational constant, M the mass of the planetary object, and r the distance from the reference point.

The Doppler frequency shift experienced by the transmitted signal upon reception can therefore be calculated as:

$$\Delta f = f_S - f_E = f_S \cdot \left(1 - \frac{f_E}{f_S}\right) = f_S \cdot \left[1 - \frac{1 - \hat{n} \cdot \vec{\beta}_E + \frac{1}{2}|\vec{\beta}_E|^2 - \Phi_E/c^2}{1 - \hat{n} \cdot \vec{\beta}_S + \frac{1}{2}|\vec{\beta}_S|^2 - \Phi_S/c^2}\right] \tag{4.17}$$

In order to perform the calculation of the vector $\vec{\beta}$ and the gravitational potential Φ, needed to determine the resulting Doppler shift, the reference system and the time basis must be carefully considered when defining the observation geometry. Each vector must be defined at the relevant time (transmission time does not coincide with reception time, which includes the calculation of the light-time) and all vectors must be defined w.r.t. the same reference system or transformed before performing vectorial summation. Furthermore, all velocity vectors must be added accordingly to relativistic summation (see par. 3.2).

4.5.1 Implementation of the "BSR Doppler Module" in the RSS

The path of reflected signal was divided into two segments: from the spacecraft to the planet surface and, after reflection, from the surface to the ground station on Earth. The specular point was considered as receiver in the first segment (called "up-link", for analogy with common G/S- to S/C transmissions) and as transmitter in the second segment (called "down-link"). The total Doppler shift was calculated as the sum of both contributions. The calculation of the frequency shift of the di-

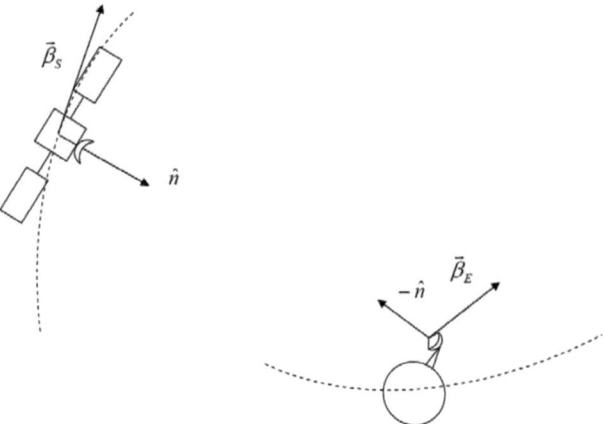

Figure 4.8: *Geometry for computation of Doppler frequency shift as in eq. (4.15).*

rect signal from the S/C was also implemented, in order to use it as a reference for the reflected signal. Fig. 4.9 shows the relevant geometry, with each vector being considered at the appropriate time. To this regard, transmission from the S/C and reflection at the specular point were considered to happen simultaneously at $t = t_0$. The assumption is justified by the small distance between S/C and planet during the experiment. As an example, an average distance of $1000\,km$ (which is quite conservative) would produce a time error of $3.\bar{3}\,ms$, which would not significantly alter position- and velocity vectors of the specular point at the surface. Similarly, reflected- and direct signal are considered to arrive simultaneously at the G/S at $t = t_1$, as the difference in the length of the traveled paths is negligible.

Considering the up-link path (from S/C to the specular point), the vectors $\vec{\beta}$ for transmitter (S/C), and receiver (specular point) are respectively calculated as (see fig. 4.9):

$$\vec{\beta}_S(t_0) = \frac{\vec{v}_{SC}(t_0)}{c} = \frac{1}{c}\frac{d}{dt}\{\overrightarrow{OP}(t) + \overrightarrow{PV}(t)\}|_{t=t_0} \tag{4.18}$$

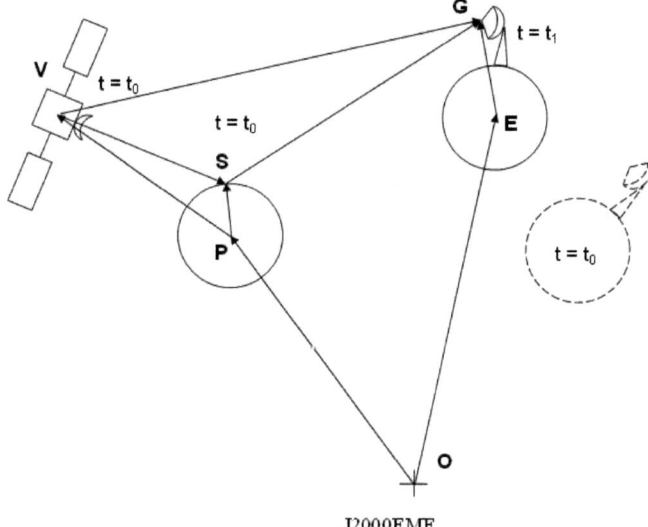

Figure 4.9: *Geometry for the computation of the Doppler frequency shift of echo- and direct signal in the BSR experiment. All vectors are expressed in J2000 EME reference system, whose origin is located at the barycenter of the solar system. All times are specified in TDB. Transmission from the S/C and reflection at the specular point are assumed to be concurrent at $t = t_0$. Similarly, reflected- and direct signal are considered to arrive simultaneously at the G/S at $t = t_1$. Dashed: Earth position at $t = t_0$.*

$$\vec{\beta}_E(t_0) = \frac{\vec{v}_{SP}(t_0)}{c} = -\frac{1}{c}\frac{d}{dt}\{\overrightarrow{OP}(t) + \overrightarrow{PS}(t)\}|_{t=t_0} \qquad (4.19)$$

The second term on the right-hand side of eq. (4.19) describes the velocity of the specular point in the inertial frame, which is due to the rotation of the planet around its axes and to the sweeping of the spacecraft antenna beam across the planetary surface. This second contribution can be computed as the time derivative of the difference of the position vectors $\Delta\overrightarrow{PS}_i = \overrightarrow{PS}_{i+1} - \overrightarrow{PS}_i$ aligned along the specular track at each considered time.

For the down-link path (from the specular point to the /GS), the vector $\vec{\beta}_S$ is the same as in eq. (4.19) (as the specular point becomes the transmitter for the D/L connection) and $\vec{\beta}_E$ is given by:

$$\vec{\beta}_E(t_1) = \frac{\vec{v}_{GS}(t_1)}{c} = \frac{1}{c}\frac{d}{dt}\{\overrightarrow{OE}(t) + \overrightarrow{EG}(t)\}|_{t=t_1} \tag{4.20}$$

Position of celestial bodies at each time is calculated by the RSS *"Ephemeris Module"*, as described in Ch. 3.

From considering eqs. (4.1), (4.2), and (4.19) follows that the spectral separation of the echo signal from the direct signal radiated via an antenna side-lobe is mainly due to the rotation of the target celestial body on its axis. In the case of the Rosetta bistatic radar experiment at the comet 67P/Churyumov-Gerasimenko, the small angular momentum of the comet will result in the spectral closeness of direct signal and echo. A coarse evaluation of the frequency shift Δf of the echo w.r.t. the direct signal is obtained by setting:

$$\Delta f = \frac{v_{tang}}{\lambda}, \tag{4.21}$$

with v_{tang} the magnitude of tangential velocity of a point on the surface, and λ the wavelength of the probing RF carrier signal. Assuming for the comet a spherical shape of radius $r = 2\,km$, and a rotation period $T = 12\,h$ (see [61]), and considering X-Band signal ($\lambda = 3.6\,cm$) yields $\Delta f \simeq 8\,Hz$.

In this case, the implementation of an ultra-stable oscillator (USO) on-board guarantees very high stability of the frequency source, which is required in order to separate spectral contributions by large integration times of the received signals.

Besides the mentioned time approximation, a second simplification was made regarding the bending effect of the Venus atmosphere on the signal. In the computation of the Doppler frequency shift (as well as in the determination of the specular point on ground) the presence of the atmosphere was not considered. The effects of the simplification are illustrated in fig. 4.10: in the hypothesis of homogeneous plane atmospheric layers, the effective outgoing ray (vector $\overrightarrow{HI}$) is parallel to the fictive one (vector $\vec{FG}$). As a consequence, the main lobe of the effective outgoing radiation is displaced from the computed one.[7] This is not of concern: considering an offset on ground of $100\,km$ (displacement between the computed specular point and the effective one, length of the segment $\overline{DH}$) and a distance Venus- Earth of $0.5\,AU$, this would result in a angular offset of ~ 0.36 seconds of arc, which is a fraction of the $3dB$−aperture of the antenna at S-Band in the magnitude order of 10^{-5}. (The same applies for X-Band, as the two D/L frequencies have the ratio 3/11.)

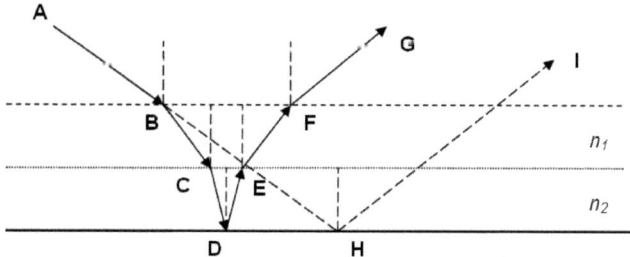

Figure 4.10: *Ray path approximation in the geometry used for the calculation of the specular point for the bistatic radar experiment at Venus. Two signal paths are displayed, with- and without consideration of atmospheric effects. Two plane homogeneous atmospheric layers with refraction index n_1 and n_2 are considered in the computation of the ray path from point A to point G through the specular point located in D. The ray path without atmospheric effects goes from point A to point I through the specular point located in H.*

After separate calculation of the $\vec{\beta}$ vectors and of the gravitational potentials for each individual signal path, up-link and down-link frequency shifts were put together in order to obtain the total Doppler frequency shift of the echo signal. The relationship (4.17) was also applied to the calculation of the frequency shift of the direct signal. From eq. (4.17) follows: $f_E = f_S - \Delta f$; by substituting $\Delta f / f_S$ with P_{UL} (for the up-link case), with P_{DL} (for the down-link case), or with $P_{DL_{dir}}$ (for the direct signal from the S/C) in the relationship (4.17), three quantities were

---

[7]The radiation arising from the reflection on ground is regarded as an antenna pattern given by the scattering function of the illuminated surface.

separately calculated:

$$f_{E_{dir}} = f_S \cdot (1 - P_{DL_{dir}}) \tag{4.22}$$

$$f_{E_{UL}} = f_S \cdot (1 - P_{UL}) \tag{4.23}$$

$$f_{E_{DL}} = f_{E_{UL}} \cdot (1 - P_{DL}) = f_S \cdot (1 - P_{UL}) \cdot (1 - P_{DL}) \tag{4.24}$$

In order to apply the relationships from (4.22) to (4.24), the frequency of the USO-driven S/C carrier, f_S, must be evaluated. This was made by interpolation over a look-up table, filled up with values from observations carried out throughout the mission during occultation seasons. Final result of computation of Doppler frequency shift of the echo signal w.r.t. direct signal, $f_{E_{DL}} - f_{E_{dir}}$, is shown in fig. 4.11. The calculation was performed for the BSR experiment of June 15, 2006, in order to compare the simulation with the data obtained from the experiment shown in fig. 4.7. Both figures show the frequency signature of the echo signal w.r.t. the direct carrier signal, which was steered to a constant frequency by the down-conversion process. The direct carrier signal is located at about $42\,kHz$ in the spectrogram of the received RCP signal (upper-left of fig. 4.7), whereas in the simulation it is centered at $0Hz$ (not shown). Furthermore, the signal tracks of the data in the spectrograms of fig. 4.7 are broken at 01:57 because of the introduced tuning offset of $29kHz$, as mentioned. The simulation shows the Doppler shift of the echo signal coming from the geometric specular point, without considering reflections from the surrounding illuminated area. As a consequence the frequency signature is represented by a line, whereas real data show dispersion around the signature of the specular point.

Comparison of fig. 4.11 with fig. 4.7 shows good agreement of the computation with the measurement.

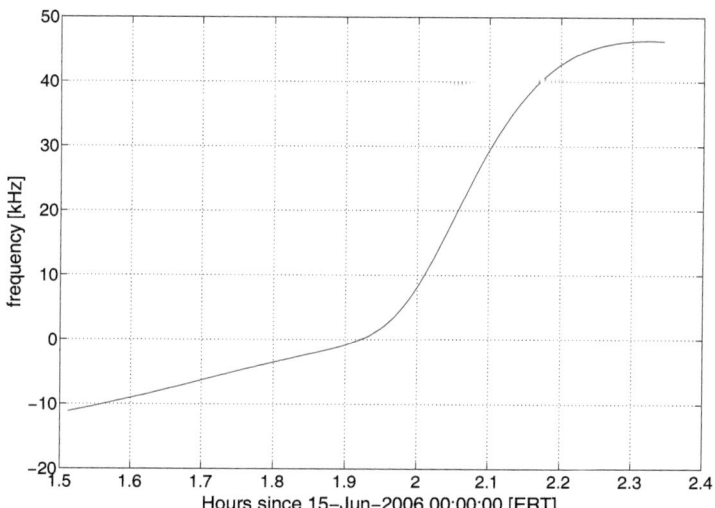

Figure 4.11: *Expected Doppler frequency shift for the VEX BSR experiment on June 15, 2006. Direct carrier signal is centered at origin of frequency axis (not shown). Simulation exhibits consistency w.r.t. data shown in fig. 4.7.*

Chapter 5

The Radio Occultation Method

5.1 The "Occultation" Experiment

The probing of planetary (or terrestrial) ionospheres and/or atmospheres by means of radio signals exploits the variation of the value of the refractive index along the signal optical path in the investigated medium (media), since it reflects itself on time-dependent changes of some signal parameters. Generally, it is possible to observe either modifications of the refractive index of the same portion of the medium, observed over time (i.e. ground-based radar observation of the terrestrial atmosphere, [84]), or a time series of consecutive observations of different regions of the investigated medium, each one characterized by a different value of the refractive index, which is assumed to be stationary; hereby the dependency of the refractive index on the position is translated into time-dependent variations, which, therefore, can be evaluated in the received probing radio signal. The latter approach is at the base of the *occultation method*, which exploits the transit of the investigated medium across the line-of-sight connecting transmitting- and receiving antenna.

Typically, the occultation geometry is a bistatic configuration, which is carried out by means of spaceborne applications for one- or both of the involved radio systems.

The VEX-*VeRa* occultation experiment (OCC) is performed in the *one-way* transmission mode (see par.2.2): the spacecraft transmits a down-link carrier signal, which, after transit in the Venus atmosphere, is received by the NNO G/S[1].

---

[1]The draw-back of the *two-way* in occultation experiments is the necessity to reestablish the radio-link coherency after the deep occultation phase, when the level of the on-board received

The occultation measurements are performed in so-called "Earth occultation" constellations. In this observation geometry, the spacecraft-to-Earth communication link is interrupted by the planet itself. Throughout the mission, the relative positions of Earth and Venus varies continuously leading periodically to so-called *"Occultation Seasons"*. Each occultation season lasts about two to three months depending on the planetary constellation, as shown in fig. 5.1. Fig. 5.2 shows the latitudinal distribution of the Earth occultation ingress and egress points as a function of time for the nominal and extended mission periods. An occultation pass can last up to 50 min. depending on the orbital/geometrical configuration among spacecraft, Venus, and Earth ([45]). The maximum occultation times occur in the middle of the occultation season, when VEX passes diametrically behind Venus and the ingress/egress points are essentially at the Venusian poles (diametric occultation). Although the coverage is best at northern polar latitudes, occultations occur at all planetocentric latitudes over both hemispheres ([46]).

In a typical occultation pass, as the satellite approaches the planet, the down-link signal from S/C to the G/S consecutively penetrates interplanetary space, planetary ionosphere, and neutral atmosphere. This represents the *"ingress"* phase of the pass. The reverse sequence, as the S/C emerges from behind the planetary disk, is called the *"egress"* phase. The interaction with the planetary media alters the transmitted radio signal by impressing time-dependent variations on its parameters. By evaluating the received signal is then possible to derive vertical profiles of planetary atmospheric- and ionospheric parameters.

During an occultation pass, the very dense atmosphere of Venus with its associated strong density gradients causes continuous refraction of the transmitted RF carrier signal. As a consequence, in order to compensate the bending angle –which would deviate the emerging signal away from the Earth– special S/C attitude maneuvers must be executed, which result in a continuous steering of the spacecraft high-gain antenna.

---

signal begins to increase over the sensitivity threshold of the receiver ($\sim -146\,dBm$ for both, X- and S-Band; narrow-band acquisition. [16]), thus missing on a portion of the measurement. On the other hand, the *one-way* radio-link presents the disadvantage of a less stable frequency source (USO vs. H_2-Maser, see par. 2.2); however, this is not of concern: even if two orders of magnitude larger than the frequency stability of the G/S maser, the frequency stability of the USO meets the specified accuracy requirements ([43]).

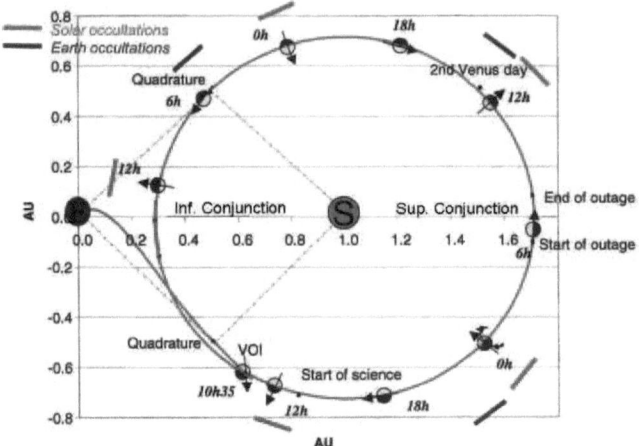

Figure 5.1: *Nominal mission overview in the Sun-Earth fixed coordinates. Arrows define orbital plane (perpendicular to the planetary disc) and direction of the S/C motion along the orbit. Red- and blue bars mark Solar- and Earth occultation, respectively. Local time at ascending node is also specified (LTAN).* ([96]).

The proper S/C attitude, which allows occultation observations and which is different for each occultation pass, is computed by the *VeRa* team by means of the *Radio Science Simulator* (RSS), a dedicated software tool which was developed by the *VeRa* team in order to support the Radio Science investigation (see Ch.3).

The physical behavior of the Venus atmosphere has also the effect that it can only be sounded to a minimum height of approximately $33\,km$, the so-called "super-refraction height". Here the curvature of the ray path approaches the curvature of the planet's surface (atmosphere). ([68], [45], [46]). (See par. 5.2).

5.2 Measurement principle

The purpose of an occultation experiment is the retrieval of the *refractivity* μ, with $\mu = (n-1) \times 10^6$, and n refractive index, as a function of the radial distance from the center of the planet, r.

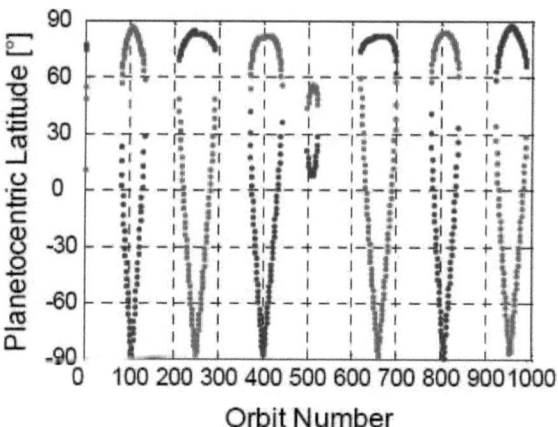

Figure 5.2: *Distribution of planetocentric latitudes for occultation ingress (blue) and egress (red) at the planetary disk for the nominal (orbits 0-550) and extended (orbits 551-1000) Venus Express mission.* ([46]).

The refractivity is a key parameter for the investigation of both, ionosphere, as it is related to the *electron number density*, and neutral atmosphere, since μ is proportional to the *number density* of the different constituents of the atmosphere (see eq. (5.16)).

As the ideal gas law relates pressure, temperature, and number density of the gas (see eq. (5.17)) and since hydrostatic equilibrium applies very accurately throughout well-mixed atmospheres, it is possible to derive the vertical structure of the neutral atmosphere from the refractivity profiles ([23], [56], [46], [94]).

Continuous refraction of the radio signal in the investigated atmosphere causes the wave vector to follow a curved trajectory in medium. If spherical symmetry is assumed, the refractive index is function only of radius r (radial distance from the origin of the planetocentric coordinate system). In this case, the experiment can be described by means of a bi-dimensional geometry, where the involved parameters are described in the plane formed by the spacecraft, the ground station, and the gravity center of the reference planet.

Figure 5.3 show the experiment geometry in a Venus-centric coordinate system (z, x), along with the relevant parameters: the *refraction angle* α, the *impact parameter* a, and the distance of *closest approach* r_0. These parameters are to be understood in the context of geometric optics, where they can be put in relationship with the perturbations of the frequency and amplitude (the observables) of the probing signal, which are caused by continuous refraction in the transected medium.

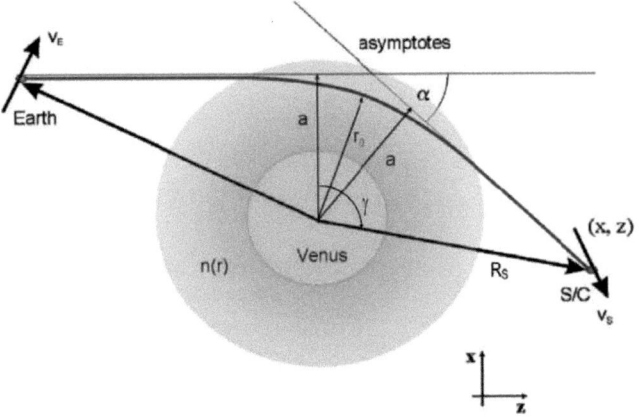

Figure 5.3: *Geometry used for calculation of an electromagnetic wave refraction in the Atmosphere of Venus during an occultation. Ray path closest approach distance r_0 and deflection angle α are related to the impact parameter $\mathbf{a}$ (asymptote closest approach distance) and index of refraction $\mathbf{n}(\mathbf{r})$. The (z,x) coordinate system is a planetocentric coordinate system. ([46])*

For radial variations on a scale large compared with the radio wavelength, the radio waves propagate locally as in a homogenous medium but follow paths that can be determined by ray optics ([23]). Referring to the geometry shown in fig. 5.3, where the (z, x) plane is the plane containing S/C, G/S, and Venus at a given moment in time, the asymptotes of the considered ray, that is, the tangents to the curved ray in the transmission- and reception directions, cross each other over the point of closest approach, given by the vector $\vec{r_0}$, forming the refraction angle α. The distance of both asymptotes from the center of the planet is the impact parameter a. Knowing position and velocity of the S/C, G/S, and planet at each time, it is possible to relate the Doppler frequency shift of the received signal to the refraction angle α and impact parameter a.

In the method proposed by Fjeldbo ([26], [27]) for the evaluation of the *Mariner V* occultation data, and re-elaborated by Eshleman ([23]), and later by Häusler ([45], [46], [47]), the measured Doppler shift is related to the projection of the spacecraft- and ground station velocities –given in the Venus reference frame– onto the ray asymptotes. Referring again to fig. 5.3, considering a stationary Earth and planet, parallel rays from the Earth, and the S/C outside the atmosphere, so that both impact parameters have the same length, the *frequency residual* Δf, which accounts for the medium-induced effects, is (approximation to first order in v/c, [52]):

$$\Delta f = \frac{1}{\lambda} \cdot [v_x \sin\alpha + v_z (1 - \cos\alpha)] \ , \tag{5.1}$$

The quantities in eq.(5.1) are time-dependent: the variations of the retrieved frequency residual reflect the variations of the refraction angle while the RF carrier signal propagates through the sounded atmosphere. From eq. (5.1) follows that the sensitivity of occultation measurements to the variations of α (minimum value of α which can be resolved) is dictated by the uncertainty of the retrieved frequency residual. Whereas the evaluation of the accuracy attainable in the computation of the frequency residuals is object of Ch. 8, a quick estimate of the experiment sensitivity is given in par. 5.2.1.

The time series of the values of α, obtained by evaluating the time-dependent frequency residual in (5.1), enter a second equation, which relates the bending angle to the impact parameter a:

$$a = R_s \cos(\gamma - \alpha) \ , \tag{5.2}$$

where R_s is the S/C distance from the origin of the reference system, and γ is the angle between the asymptote at the Earth side and R_s. The values of $\vec{v}_{S/C\,rel}$, R_s, and γ can be determined from spacecraft tracking data, station location information, and precise planetary ephemerides. Thus a measurement of Δf as a function of time establishes both α and a as functions of time. From these it is possible to form $\alpha(a)$, the ray bending angle as a function of the impact parameter ([23]).

Based on general principles of optics ([8]), Fjeldbo et al. ([27]) showed that determining the refractive index profile from $\alpha(a)$ is a special case of *Abelian* integral inversion. For a ray path through the atmosphere, it can be shown that for a radially symmetric atmosphere the asymptotic bending angle α is related to the refractive index n via an Abel transform ([27], [23], [46]):

$$\alpha(a) = -2a \int_{a/n}^{\infty} \frac{1}{n} \frac{dn}{dr} \frac{dr}{\sqrt{(nr)^2 - a^2}}, \qquad (5.3)$$

with r the radius from the planet center to the ray.

Inverting equation (5.3) yields the refractive index as a function of the atmospheric height (radius) ([27], [23], [46]):

$$\pi \ln[n(r_{01})] = -\int_{a=a_1}^{a=\infty} \frac{\alpha(a)}{\sqrt{a^2 - a_1^2}} da \qquad (5.4)$$

Eq. (5.4)[2] gives the value of the refractive index for a particular value r_{01} of the radius r as a function of a_1, which is a particular value of the impact parameter a. But, since r_{01} and a_1 are related by the Bouguers Rule ([26], [23]):

$$r_{01} = \frac{a_1}{n(r_{01})}, \qquad (5.5)$$

the refractive index can be determined as a function of the radius from the ray bending angle, which in turn is determined from the Doppler frequency measurements ([23]).

The inferior limit for the radial distance is given by the height of *critical refraction*, or *super-refraction*, when the radius of curvature of a horizontal ray is equal to the distance to the planetary center of mass ([27]). This fixes the maximum probing depth at Venus to $r - R_V \sim 33\,km$, with $R_V = 6051.8\,km$. As we will show later

---

[2] In order to avoid the singularity $a = a_1$, eq. (5.4) can be manipulated to: $\pi \ln[n(r_{01})] = -\int_{a_1}^{\infty} \frac{d\alpha(a)}{da} \cosh^{-1}\left(\frac{a}{a_1}\right) da.$ ([23]).

(see par. 7.2.1), *VeRa* can perform atmospheric sounding down to depths comprised in the range of $36-35\,km$, achieved by means of the OL data processing software.

5.2.1 Actual optical path

The "optical path" of an electromagnetic wave propagating in a medium is a fictive path, larger than the actual traveled path, as it accounts for the effect of the reduction of the propagation velocity w.r.t. vacuum.

The ratio between the speed of light in vacuum, c, and the speed of light in a medium, c', corresponds to the refractive index n of the considered medium:

$$\frac{c}{c'} = \frac{1}{\sqrt{\mu_0 \epsilon_0}} \cdot \sqrt{\mu \epsilon} = \sqrt{\mu_r \epsilon_r} = n \qquad (n > 1 \Leftrightarrow c' < c) \; , \tag{5.6}$$

therefore the optical path is calculated by "weighting" the actual path by the refractive index:

$$l_{opt} = \int_{l_m} n \, dl \tag{5.7}$$

where l_m is the curved path followed by the wave in the medium (actual path).

The frequency shift considered in eq. (5.1) is an approximation. Actually, there are two causes for the phase difference experienced by the received signal w.r.t. the vacuum, as shown by eq. (5.7): the different propagation path (which is curved and not straight), and the variations of the refractive index along this path. Therefore, a more accurate solution shall consider integrating the refractive index along the actual traveled path.

The optical path difference w.r.t. the vacuum is:

$$\Delta l = \int_{l_m} n \, dl - \int_{l_0} dl, \tag{5.8}$$

where l_0 is the straight-line path followed by the wave in vacuum. This leads to a phase difference $\Delta\phi$ which is given by:

$$\Delta\phi = \phi - \phi_0 = \frac{2\pi}{\lambda} \int_{l_m} n\, dl - \frac{2\pi}{\lambda} \int_{l_0} dl; \qquad (5.9)$$

where λ is the wavelength in vacuum. The correspondent frequency shift w.r.t. vacuum can be calculated as:

$$\Delta f = \frac{1}{2\pi} \frac{d(\Delta\phi)}{dt}. \qquad (5.10)$$

Referring to fig. 5.3, the elementary path dl is ([79]):

$$dl = \frac{dr}{\cos\alpha} \qquad (5.11)$$

with dr the elementary radial distance. From the Bouguers Rule ([26]), $a = n r \sin\alpha$, follows ([79]):

$$dl = \frac{n r}{\sqrt{n^2 r^2 - a^2}}\, dr \qquad (5.12)$$

Inserting eq. (5.12) in (5.9), considering that $n = 1$ in vacuum, yields ([79]):

$$\Delta\phi = \frac{2\pi}{\lambda} \int_{r_m} \frac{n^2 r}{\sqrt{n^2 r^2 - a^2}}\, dr - \frac{2\pi}{\lambda} \int_{r_0} \frac{r}{\sqrt{r^2 - a^2}}\, dr \qquad (5.13)$$

The approximated expression of the frequency residual shown in eq. (5.1) can be derived from the exact relationship (5.13) under the assumption of small changes of the optical path in the atmosphere ([79]). Applying eq. (5.10) to eq.(5.13) yields for the frequency contribution of the atmosphere:

$$\Delta f = \frac{1}{2\pi} \frac{d(\Delta\phi)}{dt} \sim \frac{v_x \sin\alpha}{\lambda} + \frac{v_z (1 - \cos\alpha)}{\lambda} \qquad (5.14)$$

which corresponds to eq.(5.1).[3]

Experiment sensitivity

As already mentioned in par. 5.2, the sensitivity of the experiment depends on the sensitivity to the variations of the refraction angle α, which is dictated by the accuracy attained in the calculation of the frequency residuals.

For very small values of the bending angle near the top of the atmosphere/ionosphere, at X-band frequencies ($\lambda = 3.6\,cm$), assuming in (5.1) $v_x \simeq 8\,km/s$, and $v_z \simeq 0$, and a residual frequency noise of the order of $1\,mHz$ (see Ch. 8), we obtain for the angle increment $\Delta \alpha$:

$$\Delta \alpha \simeq 5 \times 10^{-9} \quad (rad) , \qquad (5.15)$$

(see also [46]). Near the top of the atmosphere/ionosphere, based on an observation interval of $1\,s$, this deflection angle can be transformed via an Abel transform into a refractivity error of the order of $\Delta \mu \leq 10^3$, ([46]).

5.2.2 Atmospheric Profiles

The refractivity profile $\mu(h)$ as a function of the altitude h is determined by the local state of the neutral atmosphere and the electron density distribution $N_e(h)$ in the ionosphere as ([94]):

$$\mu(h) = -C_3 \cdot \frac{N_e(h)}{f_0^2} + C_1 \, n(h) \, k, \qquad (5.16)$$

where k is the Boltzmann constant ($1.38065(26) \times 10^{23}[J/K^{-1}]$, [70]), $n(h)$ is the vertical distribution of neutral number density, $C_3 = 40.31[m^3/s^2]$, and f_o is the frequency of the radio signal. The first term in eq. (5.16) describes the effect of the ionospheric electron density. The neutral number density is related to the refractivity profile through the constant factor C_1, which depends on the composition of the

---
[3]Relativistic effects are being treated in [47], including also effects from the General Theory of Relativity.

atmosphere. This proportionality factor has the value $C_1 \sim 1312 \times 10^{-6} [Kms^2/kg]$ for an atmospheric composition of 96.5% CO_2 and 3.5% N_2 ([24], [94]). At Venus, the ionospheric contribution is much weaker than the one from the neutral atmosphere. It has been observed from the evaluation of *closed-loop* data from VEX occultation passes that the neutral and ionized regions are well separated in altitude, the atmospheric term dominating below about $100\,km$, the ionosphere above $100\,km$ ([72], [94]).

Pressure, temperature, and neutral number density of an atmosphere are related through the ideal gas law ([94]):

$$p(h) = k \cdot n(h) \cdot T(h) \tag{5.17}$$

Since hydrostatic equilibrium is very accurate throughout a well-mixed planetary atmosphere, height profiles of pressure and temperature can be derived directly from the neutral number density profile ([23], [56], [46], [94]). The temperature is calculated as:

$$T(h) = \frac{\mu_{up}}{\mu(h)} \cdot T_{up} + \frac{\bar{m}}{k \cdot n(h)} \int_h^{h_{up}} n(h') \cdot g(h') \, dh', \tag{5.18}$$

where $\bar{m}$ represents the mean molecular mass of the mixed neutral atmospheric species, and $g(h)$ the altitude-dependent acceleration of gravity ([64]). The integration parameter $T_{up} = T(h_{up})$ represents the temperature at the upper boundary of the detectable atmosphere.

The pressure vs. height profile is given by:

$$p(h) = \frac{1}{C_1} \mu(h) \, T(h). \tag{5.19}$$

Fig. 5.4 show a typical temperature profile, retrieved from CL data obtained during the VEX-VeRa occultation pass on DoY234-2006, orbit #123, at latitude $71°\,N$, derived with three different upper boundary temperature conditions of 170, 200, and $230\,K$. Regardless of the upper boundary condition, all three profiles converge to the same temperature distribution below $90\,km$. The temperature shows a constant cooling (lapse rate) of $-10\,K/km$ within the cloud deck below $60\,km$. The inversion at $62\,km$ marks the tropopause, the transistion from the troposphere to the mesosphere. On average, the temperature is isothermal $(233\,K)$ within the upper cloud deck up to $75\,km$, but displays significant fine structure, a series of small inversions, with amplitudes larger than the measurement error ([72]).

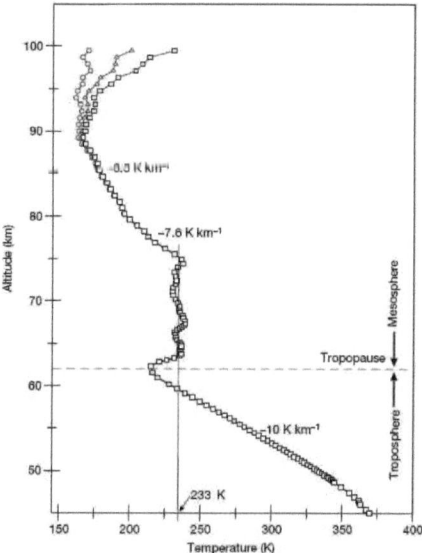

Figure 5.4: *Typical temperature profile, retrieved from CL data obtained during the VEX-VeRa occultation pass on DoY234-2006, orbit #123, at latitude $71°\,N$, derived with three different upper boundary temperature conditions of 170, 200, and $230\,K$ ([72]). (Comments: see text).*

Fig. 5.5 shows a temperature map based on 12 temperature profiles from the southern hemisphere obtained during the third occultation season during nighttime from DoY 116-2007, to DoY 146-2007). Some typical characteristic of the Venusian atmosphere are apparent in this map. The most striking feature is the presence of a strong inversion layer at a pressure level near $100\,hPa$ ($\simeq 60-70\,km$) that becomes apparent in the latitude range between $60°$- and $80°$ latitude. This inversion layer corresponds to the "cold collar" region (see [94]).

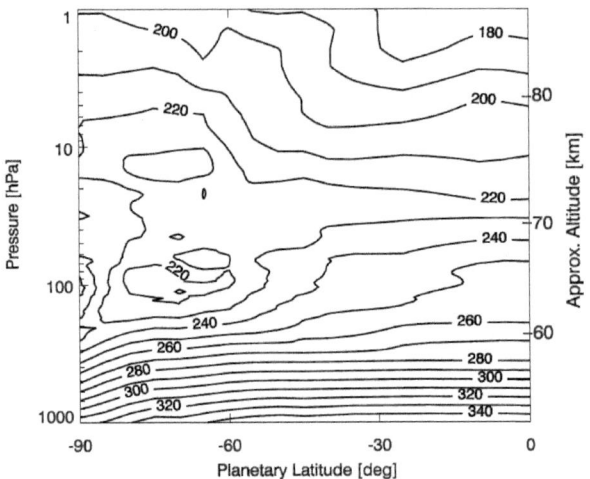

Figure 5.5: *Nighttime temperature in the southern hemisphere for the third occultation season. Local times range from 22:30 at low latitudes to 05:00 at high latitudes. Contour map is based on measurements from 12 occultations in the interval DoY 116-146, 2007* ([94]).

5.2.3 Signal Attenuation: Absorption and Defocusing-Loss

Besides refraction effects, a planetary atmosphere of neutral gaseous molecules and atoms exhibits also absorptive radio effects ([23]). Both affect the amplitude of the received signal. Since the effects of refraction are contained also in the frequency signature of the received signal, it is possible to calibrate them out in order to compute absorption profiles as function of altitude from the observed signal attenuation.

Defocusing-Loss

The gradient of the refractive index with respect to the radius and the assumed spherical symmetry spread rays forming the antenna beam in the plane of occultation. By defocalizing the probing signal, the atmosphere acts as a diverging lens. The resulting effect is an attenuation τ of the received signal w.r.t. the vacuum case, the *defocussing loss*, which can be calculated as ([23], [79], [46]):

$$\tau_{|dB} = 10\, Log\left(cos\alpha - D\frac{d\alpha}{da}\right). \tag{5.20}$$

where D is the distance from the spacecraft to the crossing of the asymptotes. Therefore once the refraction angle as function of the impact parameter has been retrieved ($\alpha(a)$), it is possible to calculate the defocussing loss and subtract this quantity from the total attenuation of the received signal in order to estimate the *dispersive absorption* in the atmosphere.

Eq. (5.20) is an approximate expression. A more accurate analysis of defocusing effects is afforded in ([64]). Here, besides the energy spreading in the plane of refraction, another effect is considered: the compression of the rays in the plane perpendicular to plane of refraction. Fig. 5.6 shows that the signal energy crossing the surface of the annulus of inner radius a is distributed on the annulus of inner radius r. In order to calculate the effect of the phenomenon on the received signal strength, a correction factor has been considered, which corresponds to the ratio of the two radii: $\phi_1 = a/r$. Depending on which radius is larger, ϕ_1 accounts for a loss or for a gain. The analytic expression of ϕ_1 is: ([64])

$$\phi_1 = \left(\frac{1}{cos\alpha} - \frac{D}{a}\frac{sin\alpha}{cos\alpha}\right)^{-1} \tag{5.21}$$

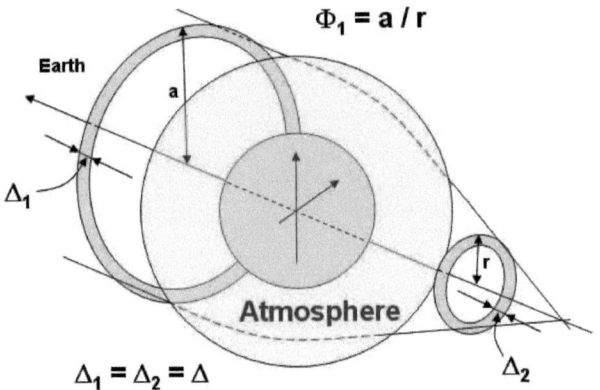

Figure 5.6: *Geometry for the computation of the defocussing loss factor ϕ_1* ([81]).

In order to account also for the effects of defocusing in the plane of refraction (that is, the effects of the differential refraction of the rays contained in the antenna beam), a second corrective factor, ϕ_2, is considered in [64]. Fig. 5.7 shows that the distance Δa vertically separating incident rays is spread over the distance Δr after refraction.[4] For ϕ_2 is: ([81])

$$\phi_2 = \left(\frac{\partial r}{\partial a}\right)^{-1} = \left[1 + \left(a\, tg\alpha - \frac{D}{\cos\alpha}\right)\frac{\partial \alpha}{\partial a}\right]^{-1} \qquad (5.22)$$

The overall defocussing-loss-factor is, therefore:

$$\phi_1 \cdot \phi_2 = \cos\alpha \cdot \left(1 - \frac{D}{a}\sin\alpha\right)^{-1} \cdot \left[1 + \left(a\, tg\alpha - \frac{D}{\cos\alpha}\right)\frac{\partial \alpha}{\partial a}\right]^{-1} \qquad (5.23)$$

---

[4]Fig. 5.7 considers an *up-link* configuration: incident, parallel rays transmitted by the G/S (left-hand side) are received by the S/C (right-hand side) after differential refraction in the atmosphere. The considered geometry applies also for the down-link case, which is obtained by swapping the positions of S/C and G/S in the figure. In this case, the transmitted incident rays can still be considered parallel, since the distance of the S/C is much larger than the distance between rays.

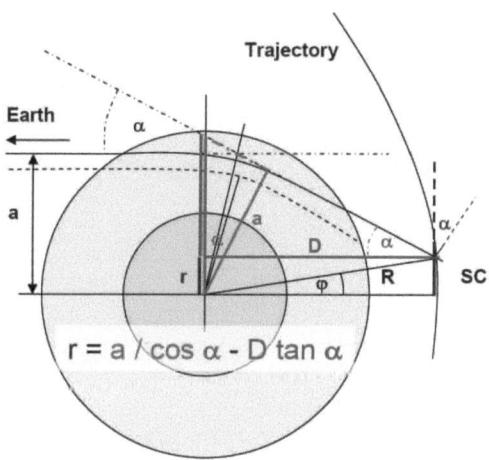

Figure 5.7: *Geometry for the computation of the defocussing loss factor ϕ_2 ([81]).*

By means of the retrieved values of refraction angle α, impact parameter a, and distance between S/C and asymptotes crossing point D, a profile of the defocusing loss as function of the height above ground can be calculated by means of eq. (5.20), or eq. (5.23).

Absorption

In order to estimate absorption effects in the investigated atmosphere, the computed defocusing loss profile must be subtracted from the attenuation curve of the received signal. Fig. 5.8 shows the results obtained from CL data collected during the VEX-VeRa occultation pass on DoY196-2006, orbit #85.

The measured signal attenuation (red curve) can be compared with predicted defocusing loss (blue curve, dashed), and computed defocusing loss (blue curve, continuous) from formula (5.20). The difference between the measured attenuation and the defocusing loss yields the absorption profile (green curve, continuous).

5.2.4 Spatial Resolution

In the adopted geometric optics approach no diffraction effects are considered. Therefore the spatial resolution of the atmospheric probing is given by the exten-

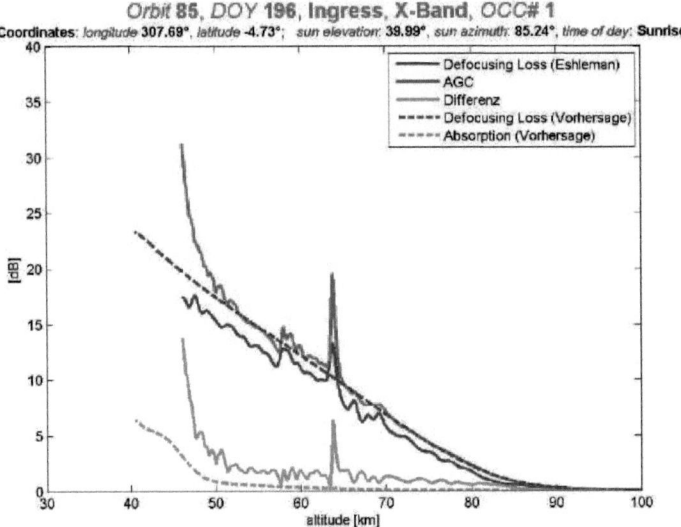

Figure 5.8: *Comparison of measured signal attenuation (red curve) with predicted- (blue curve, dashed), and computed (blue curve, continuous) defocusing loss from formula (5.20). The difference between the measured attenuation and the defocusing loss yields the absorption profile (green curve, continuous), whereas the dashed green curve represents a model-based prediction of the absorption loss ([60]).*

sion of the first *Fresnel zone* in the crossed medium. Due to the assumed spherical symmetry (radial refraction gradient), the Fresnel zone is flattened in the plane of refraction and undisturbed from its free space value perpendicular to this plane, that is, the effect of refraction is that of changing the section shape of the Fresnel zone from circular to elliptical. The result is an improved altitude resolution in regions for which defocusing is larger.

The horizontal- and vertical radius of the Fresnel zone, r_H and r_V, respectively, are ([46]):

$$r_H = \sqrt{\frac{\lambda\, db}{d+b}}, \qquad (5.24)$$

$$r_V = r_H \sqrt{\frac{I_{refr}}{I_0}}, \qquad (5.25)$$

where λ is the wavelength, d is the distance from the transmitter (S/C) to the atmospheric reference point, and b is the distance from the planet to the receiver (G/S). The second factor on the right-hand side of eq. (5.25) represents the defocusing loss (see par. 5.2.3). Usually is $d << b$, so that eq. (5.24) can be approximated by:

$$r_H \sim \sqrt{\lambda d}. \qquad (5.26)$$

For X-band frequencies ($\lambda \sim 3.6\,cm$) and for typical distances ($d = 10.000\,km$), the horizontal diameter of the Fresnel zone –which is approximately the horizontal resolution– is $2\,r_H \sim 1200\,m$. Considering a defocusing loss of $25\,dB$ at a sounding altitude of $\sim 40\,km$, applying eq. (5.25) leads to values of the vertical resolutions at low altitudes $2\,r_V \sim 66\,m$ ([46]). Special techniques, such as "wave back-propagation" have been developed to recover the complete refractivity profile at resolutions an order-of-magnitude better than the Fresnel scale ([50]). The techniques also provide a means of deciphering multipath propagation ([57], [52]; see par. 7.2.2).

5.3 Data kind and data acquisition

The primary observables in an occultation experiment are the received signal phase and amplitude. Two different techniques are used for the signal acquisition on ground: the "Closed Loop" (CL) mode and the "Open Loop" (OL) mode (see figs. 5.9, and 5.10).

The CL receiving system makes use of a *Phase-Lock-Loop* (PLL) circuit in order to track the incoming RF radio signal. The phase of the received signal can be directly recovered from the error signal which drives the local oscillator (LO) of the PLL, whereas its squared amplitude (peak power) is recorded by an Automatic-

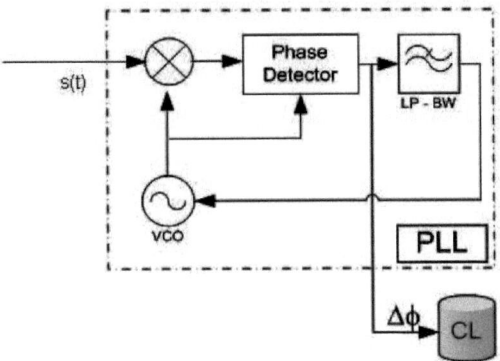

Figure 5.9: *Principle scheme of a "Closed Loop" receiver. The elements shown are defined in the text.*

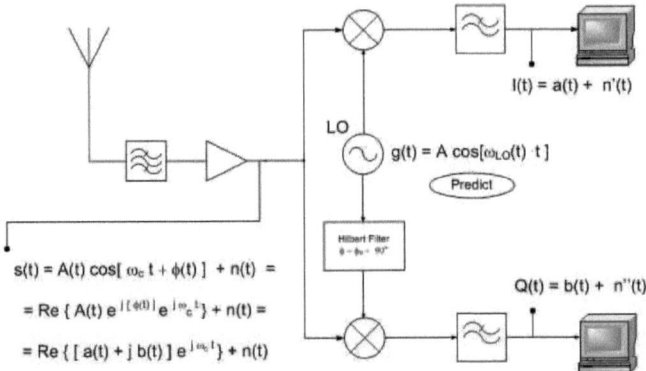

Figure 5.10: *Principle scheme of a "complex" coherent receiver. The quantities shown are defined in the text. ([68]).*

Gain-Control (AGC) circuitry.[5] The inherent characteristic of a closed-loop tracking

---

[5]The LO of the PLL is called "VCO" (Voltage-Controlled-Oscillator): it is steered by a voltage

system on the one hand makes it possible to extract information from the incoming RF signal in real-time but on the other hand makes the CL receiving technique not suitable for low sounding altitudes, which are characterized by large values of attenuation (absorption and defocusing), and signal dynamics. In fact, the PLL is not capable of maintaining the "lock-status" when the degradation of the SNR and the shift rate of the signal frequency exceed the constraints imposed by the system bandwidth (BW) of the PLL.[6] (Attenuations $> 50\,dB$ and frequency shift rates up to $2\,kHz/s$ are expected for the received signal as atmospheric-induced effects for deep occultation passes in the case of the *VeRa* experiment.)

These limitations are overcome in the OL mode by avoiding the tracking process at all. The incoming RF signal is simply heterodyned to base-band by feeding the LO with the nominal carrier frequency and prediction of the expected motion-induced frequency shift. A proper setting of the system bandwidth and sampling rate must guarantee the recovery of the full signal dynamics: the looked-for information –no longer available in real-time, as it is in the CL receiving mode– shall be contained in the recorded samples, even when masked by thermal noise. The advantage of the OL system w.r.t. CL is to have signal recorded through the whole occultation pass, which implies the possibility to extract the information from noise by application of suitable post-processing techniques.

Similarly to the case of Bistatic Radar, it is evident that numeric simulations are of central importance also for the Occultation experiment, as the settings of the receiver system at the G/S are based on the prediction of the expected frequency and amplitude of the incoming signal.

In order to recover both amplitude and phase of the base-band signal, *coherent reception* is implemented (see fig. 5.10). This kind of approach can be illustrated

---

error signal which arises from the comparison of the phases between incoming signal and oscillator signal. In order to reduce the dynamics of the error signal, the VCO is provided with the information about the expected straight-line Doppler frequency shift (the frequency shift induced by the relative motion between S/C and G/S).

[6] Signal attenuation and signal dynamics impose conflicting constraints on the system bandwidth: on the one hand the BW should be reduced to maintain a given value of the SNR when the incoming signal starts to fade out; on the other hand it should be increased when the shift rate of the frequency of received signals increases. Since signal attenuation and signal dynamics boost happen simultaneously as soon as the spacecraft signal enters the investigated atmosphere, it is not possible to meet a trade-off in the settings of the receiver.

starting from an analog signal represented in complex form:

$$s(t) = \Re\{s_c(t)\} = \Re\{A(t)\, e^{j\phi(t)}\, e^{j\omega_c(t)}\} =$$
$$= \Re\{A(t)\cos[\omega_c(t) + \phi(t)] + j\, A(t)\sin[\omega_c(t) + \phi(t)]\} \quad (5.27)$$

where $s(t)$ is the incoming radio signal, ω_c the carrier angular frequency, and $A(t)$, and $\phi(t)$ the looked-for observables of the experiment, that is, the time-varying amplitude and phase of the base-band signal, respectively. The complex signal associated to the real signal $s(t)$ is called the "analytic signal"[7]:

$$s^+(t) = A(t)\, e^{j\phi(t)}\, e^{j\omega_c(t)}. \quad (5.28)$$

The base-band component of the analytic signal is:

$$\underline{s}(t) = A(t)\, e^{j\phi(t)}, \quad (5.29)$$

which in Cartesian form is:

$$\underline{s}(t) = a(t) + j\, b(t), \quad (5.30)$$

where the relationships hold:

$$\begin{cases} A(t) = \sqrt{a^2(t) + b^2(t)} \\ \phi(t) = \arctan\left[\frac{b(t)}{a(t)}\right] \end{cases} \quad (5.31)$$

The real- and imaginary parts of $\underline{s}(t)$, i.e. $a(t)$ and $b(t)$, are called the "base-band analogue components" of the RF signal. They are also known as the "in-phase", and the "quadrature" components, respectively, as they differ by a constant phase off-set of 90°.

---

[7]The apex "+" is due to the fact that the spectrum of such a signal has only positive frequency components (see par. 6.1.7).

The down-conversion which removes the carrier frequency is carried out by splitting the incoming RF signal into two paths and multiplying each of them by LO signals which have the same frequency but a phase off-set of 90°. The product shifts the spectrum of the RF signal by an amount which equals the frequency of the LO signal (i.e. f_{LO}) in both directions of the frequency axis. Low-pass filtering allows recovering the lower replica under the assumption that the bandwidth of the RF signal is less than the LO frequency, that is: $BW < f_{LO}$. Eventually, both base-band analogue components of the RF signal are obtained as output of the low-pass filtering process of each channel. They can be digitized and stored separately.

Actually, coherent sampling is achieved at the NNO G/S by a different implementation: the signal is digitized directly after down-conversion to IF. A sampling rate (slightly larger than) twice the Nyquist bandwidth guarantees that each second sample belongs to the quadrature channel. Further operations after analog-to-digital conversion are merely numeric: the samples are down-converted to base-band by removing the IF frequency and the predicted straight-line Doppler frequency shift, filtered, decimated, and stored (see fig. 5.11). Eventually, the OL raw data consist in two sets of orthogonal samples, $I[n]$ and $Q[n]$, the *in-phase* sequence, and the *quadrature* sequence, respectively. (Brackets represent index notation, as usual for discrete-time signals.) The implemented sampling rate is $100 ksample/s$, which allows to cope with signal dynamics less smaller than $50kHz$.

As in the case of analog signals, both sequences can be combined in the complex notation:

$$s^+[n] = I[n] + j\,Q[n] = A[n]\,e^{j\phi[n]} \tag{5.32}$$

where:

$$\begin{cases} A[n] = \sqrt{I^2[n] + Q^2[n]} \\ \phi[n] = \arctan\left(\frac{Q[n]}{I[n]}\right) \end{cases} \tag{5.33}$$

and $s^+[n]$ is the *analytic sequence* associated to $s[n]$, with $s[n] = A[n]\,cos(\omega_c\,n + \phi[n])$ (see eqs. from (6.49) to (7.6)). The input data to the *open loop* software is represented by the sequence $s^+[n]$ expressed by eq. (5.32).

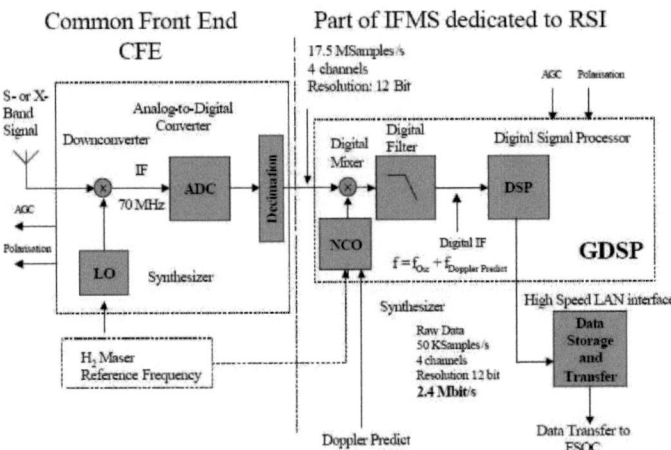

Figure 5.11: *Block scheme of the ESA New Norcia IFMS (Intermediate Frequency and Modem System) Open Loop receiver.* ([1])

Chapter 6

Elements of Signal Processing

The purpose of this chapter is to provide the reader with a short overview of the theoretic concepts exploited in the development of the *open loop* data evaluation software (OL SW).

Basic notions of signal theory, relevant for the evaluation of the data –such as linear systems, spectral analysis and filtering, modulation, Hilbert transform– and their counterpart in the theory of discrete-time signals, are introduced respectively in paragraphs 6.1, and 6.2).

The need for a clear understanding of the above mentioned concepts results evident by considering that:

- the information to be extracted from the received signal presents itself as a narrow-band[1] signal modulating amplitude and phase of the received S/C RF carrier ($\rightarrow$ *modulation*);

- in carrying out mathematical operations on the signals, the processing SW exploits the complex signal notation ($\rightarrow$ *Hilbert transform*); the availability of the information needed for complex representation is guaranteed by the implementation of *coherent reception* at the G/S ($\rightarrow$ *linear systems*);

- the signal is down-converted to base-band ($\rightarrow$ *modulation*) and filtered ($\rightarrow$ *spectral analysis* and *filtering*);

---
[1]Here the *fractional* bandwidth is intended, that is, the ratio of the bandwidth of base-band signal to the carrier frequency.

- the signal is *digitized* after down-conversion to IF (see par. 5.3), so that the input for the OL SW is represented by numeric sequences ($\rightarrow$ *discrete-time signals*, treated in par. 6.2)

The unavoidable circumstance that the received signal is corrupted with receiver- and back-ground thermal noise makes it sensible to refer to the fundamental concepts of *probability* and *stochastic processes* (treated in App. F) in order to characterize the noise process and its effect.

Finally, since *error analysis* is performed in order to validate the results (see Ch. 8), some basic notions of the theory of the *estimation* and *spectral estimation* are given in par. 6.3.

The consideration of the mentioned topics is only functional to their application to the OL SW: the treatment of the covered subjects does not have the claim of being exhaustive; extensive documentation can be found in the specified literature.

6.1 Analog signals

In the frame of the *information theory*, a signal is a quantity which is intended to convey information, typically as a function of time. It is common use to refer to a signal as a *voltage* varying over *time*, which can be expressed by the function $s(t)$[2].

6.1.1 Finite-Energy Signals and Finite-Power Signals

A signal $s(t)$ is said to be a *finite-energy signal* if the *energy* associated with it is a finite number, that is $\exists \, \mathscr{E} \in \mathbb{R} \setminus [0 < \mathscr{E} < +\infty[$. It is:

$$\mathscr{E} = \lim_{\Delta t \to \infty} \int_{-\Delta t/2}^{\Delta t/2} |s(t)|^2 \, dt \qquad (6.1)$$

A signal $s(t)$ is said to be a *finite-power signal* if the *power* associated with it is

---

[2] The mathematical function representing a signal can be either real or complex. In the latter case, the signal is called the *analytical* signal associated with the real signal (see after).

a finite number (its energy if infinite), that is $\exists\, \mathscr{P} \in \mathbb{R} \setminus [0 < \mathscr{P} < +\infty[$. It is:

$$\mathscr{P} = \lim_{\Delta t \to \infty} \frac{1}{\Delta t} \int_{-\Delta t/2}^{\Delta t/2} |s(t)|^2\, dt \tag{6.2}$$

Periodic signals of period T are comprised in the category of power signals; the associated power is:

$$\mathscr{P} = \frac{1}{T} \int_{-T/2}^{T/2} |s(t)|^2\, dt \tag{6.3}$$

6.1.2 The Fourier Series

In 1822 Jean Baptiste Joseph Fourier claimed, in the context of his work on heat flow, *"Théorie analytique de la chaleur"*, that any function of a variable, whether continuous or discontinuous, can be expanded in a series of sines of multiples of the variable. Though this result is not correct, Fourier's observation that some discontinuous functions are the sum of infinite series was a breakthrough. "The role of Fourier's thesis in signal processing comes very close to that of the *creation* of the entire discipline" ([4]). This enormous importance derives first of all from the property of linear, time-invariant (LTI) systems to respond with harmonics to harmonic inputs: a sinusoidal excitation with certain values of frequency, amplitude, and phase causes the system to respond with a sinusoidal output at the same frequency and generally different amplitude and phase; that is, harmonic signals are *eigenfunctions* of LTI systems. To characterize a system means to know its *frequency response*, i.e. the variations of amplitude and phase that the system impresses onto input sinusoidal signals characterized by different values of the frequency. When the frequency response is known, one can predict an output, as long as the input is a sinusoidal signal. Or as long as it can be seen as a *superimposition of sinusoidal functions at different frequencies*, that is, it can be expanded in Fourier's series.

The prerequisite a real function of real variable $s(t)$ must have in order to be approximated by an infinite summation of periodic functions (sinusoidal functions) is to be itself periodic. If a periodic function $s(t)$ of period T satisfies the *Dirichlet* conditions:[3]

- the integral $\int_{-T/2}^{T/2} s(t)dt$ exists and is limited;
- $s(t)$ has a finite number of discontinuities in the interval T;
- $s(t)$ has a finite number of stationary points in the interval T;

it can be expanded in a series of periodic functions, that is, the series converges to the value of the function in each point of the function's domain. At each point of discontinuity t_0, the function $s(t)$ is assigned the value: ([4])

$$s(t_0) = \frac{s(t_0^-) + s(t_0^+)}{2}, \tag{6.4}$$

where

$$s(t_0^-) = \lim_{t \to t_0^-} s(t), \tag{6.5}$$

and

$$s(t_0^+) = \lim_{t \to t_0^+} s(t). \tag{6.6}$$

The Fourier's series expansion of $s(t)$ is: ([4])

$$s(t) = \frac{a_0}{2} + \sum_{k=1}^{\infty} a_k \cdot \cos\left(2\pi \frac{k}{T} t\right) + \sum_{k=1}^{\infty} b_k \cdot \sin\left(2\pi \frac{k}{T} t\right), \quad (k \in \mathbb{Z}^+) \tag{6.7}$$

where

$$a_k = \frac{2}{T} \int_{-T/2}^{T/2} s(t) \cos\left(2\pi \frac{k}{T} t\right) dt; \tag{6.8}$$

---

[3]The Dirichlet conditions, dating back to 1837, are *sufficient* conditions: it was only in 1966 that Carleson finally discovered necessary and sufficient conditions for the convergence of the Fourier series. The topic is of a purely mathematical nature and will not be discussed here. The D. conditions are met without effort in all practical applications.

$$b_k = \frac{2}{T} \int_{-T/2}^{T/2} s(t) \, sin\left(2\pi \frac{k}{T} t\right) dt. \tag{6.9}$$

The frequencies of the trigonometric functions in eq. (6.7) depend on the period T: the *fundamental frequency* is defined as $f_0 = 1/T$; the various frequencies entering eq.(6.7) are kf_0, that is, *multiple* of the fundamental.

Equations (6.8) and (6.9) clearly show the meaning of the coefficients a_k and b_k: they express the degree of similarity of the function $s(t)$ with the harmonic functions being used in the sum and act as weights. Furthermore, it is possible to see that the cosine series approximates the *even* part of $s(t)$, whereas the sine series approximate the *odd* component.

Equations (6.8) and (6.9) are said to be the *analysis* formulas (the function $s(t)$ is *decomposed* in its harmonic components), whereas eq.(6.7) is called the *synthesis* formula (the original function $s(t)$ can be *synthesized* by adding up all harmonics (see fig. 6.1); the coefficient a_0 is an off-set, or pedestal, which plays the role of the *direct current* (DC) component: it represents the *mean value* of the signal.

Upon making use of the *Euler*'s formula:

$$e^{j\phi} = cos(\phi) + j\,sin(\phi); \tag{6.10}$$

which implies:

$$cos(\phi) = \frac{e^{j\phi} + e^{-j\phi}}{2}; \tag{6.11}$$

$$sin(\phi) = \frac{e^{j\phi} - e^{-j\phi}}{2j}; \tag{6.12}$$

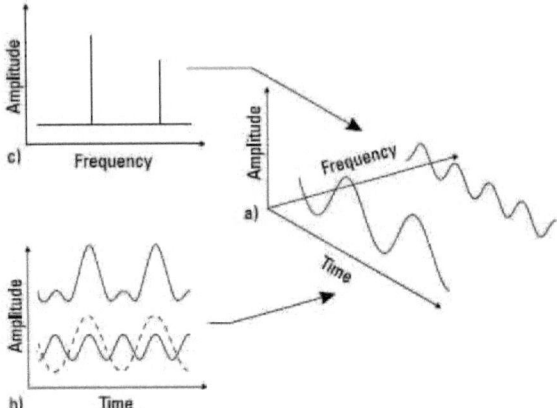

Figure 6.1: *Analysis and synthesis of waveforms by means of Fourier decomposition: the signal in* b) *(upper) consists of the sum of two sinusoids with different frequencies, amplitudes, and phases (lower). They are represented in the three-dimensional graph in* a)*, where the third dimension is given by the frequency. The graph in* c) *shows each component as a line placed at the frequency of the relevant sinusoid, whose length is proportional to the amplitude of the sinusoid.* ([2])

it is possible to come up to the *complex* form of the Fourier series:[4]

$$s(t) = \sum_{k=-\infty}^{\infty} c_k \, e^{j2\pi \frac{k}{T} t} \qquad (6.13)$$

where

$$c_k = \frac{a_k - jb_k}{2} = \frac{1}{T} \int_{-T/2}^{T/2} s(t) \, e^{-j2\pi \frac{k}{T} t} \, dt \qquad (k \in \mathbb{Z}) \qquad (6.14)$$

---

[4]Signal theory extensively avails itself of complex numbers. Signals are said to be represented in the *complex form*, or *complex notation*. The motivation lies in the Fourier decomposition: if a signal can be represented by a series of sinusoidal functions, two informations are needed: their *amplitude* and their *phase*. Polar representation of complex numbers links the information in a single number. Furthermore, linear operations among signals are more easily carried out in complex notation.

The information about how to "reconstruct" the original function $s(t)$ is contained in the complex quantities $c_k = |c_k| e^{j\varphi_k}$; they indicate how much of each component has to be taken (amplitude of the coefficient) and at which point it has to start in order to synthesize $s(t)$ (phase of the coefficient). The Fourier's coefficients c_k are functions of kf_0 and represent the *frequency spectrum* of $s(t)$; the plot of $|c_k|$ against kf_0 is called the *amplitude spectrum* (see fig. 6.3), and the plot of $\angle c_k = \varphi_k$ vs. kf_0 is called the *phase spectrum*.

Neither negative values of k ("negative frequencies"), nor the complex imaginary unit $j = \sqrt{(-1)}$ should be of concern: the complex notation is a mathematical representation of physical quantities. Even if represented as a sum of complex numbers, the function $s(t)$ in eq. (6.13) still remains a real number, representative of a physical quantity, with "physical" sinusoidal signals (therefore with *positive* frequencies!) involved in its decomposition.

An important relationship holds for the series coefficients, which goes under the name of Parseval's theorem: ([4])

$$\sum_{k=-\infty}^{\infty} |c_k|^2 = \frac{1}{T} \int_{-T/2}^{T/2} [s(t)]^2 \, dt \qquad (6.15)$$

If signals are defined as a voltage ([V]) across a resistor of value $1\,\Omega$, eq. (6.15) gives the average power as the sum of the squared amplitudes of the Fourier coefficients. Therefore, the plot of $|c_k|^2 = c_k \cdot c_k^*$ (asterisk denotes complex conjugate) against kf_0 is called the *power spectrum* of $s(t)$ ([4]).

Figs. 6.2 illustrates the application of the *analysis*- and *synthesis* formulas (see eqs. (6.8)-(6.9), and (6.7), respectively) to the case of a periodic rectangular pulse signal. The time function is decomposed into a certain number of harmonics (analysis); the components are then added together (synthesis) in order to restore the original function. The more terms are involved, the better the original function is approximated. The first term, given by the coefficient a_0 in eq. (6.7), corresponds to the DC component and it is responsible for repositioning the sum of the other terms around the mean value of the signal. Each intervening harmonic refines the approximation by "correcting" the previous term.

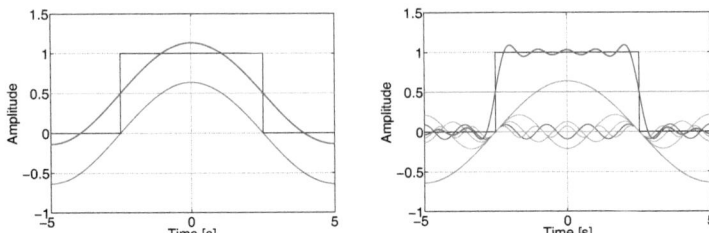

Figure 6.2: *Example of analysis and synthesis of waveforms by means of Fourier decomposition. Left-hand side: the original time function (rectangular pulse, black) is approximated by means of the first two terms of the Fourier series expansion. The first term (pink), given by the zeroth coefficient in eq. (6.7), corresponds to the DC component, or mean value of the signal and it is responsible of repositioning the sum of the other terms. The second term (blue) is obtained by considering the (6.7) for $k = 1$; it corresponds to the basic frequency $1/T$, where T is the time interval considered for the expansion. The approximating function (red), given by the summation of the harmonic components, as in eq. (6.7), is quite poor when only two terms are considered. Right-hand side: the original time function (rectangular pulse, black) is approximated by means of the first ten terms of the Fourier series expansion. The form of the underlying pulse can be clearly recognized*

It has to be noticed that, due to the even symmetry of the chosen rectangular pulse function, only the even part of the expansion is involved, that is, the cosine terms. Therefore, from eq. (6.14) follows that the coefficients of the expansion given in eq. (6.14) are real. Fig. 6.3 shows the amplitude spectrum of the complex Fourier coefficients . It can be noticed that, besides the DC component, only the odd harmonics enter the summation.

By looking at fig. 6.3 it is possible to recognize that the envelope of the amplitude spectrum follows a $sinx/x$ function, or $sinc(x)$. This particular function is the *Fourier transform* (see after) of a generic rectangular pulse function $A\,rect_\tau(t)$ (where A is the pulse amplitude and τ is the pulse duration), which represents the basic pattern of the considered periodic pulse. As a matter of fact, the Fourier coefficients of the expansion of a periodic signal represent *samples* of the Fourier transform of the base function contained in each period (see par. 6.1.3).

The spacing between the samples (Fourier coefficients) can be computed by considering that each frequency entering the expansion (6.7) is a multiple of the basic

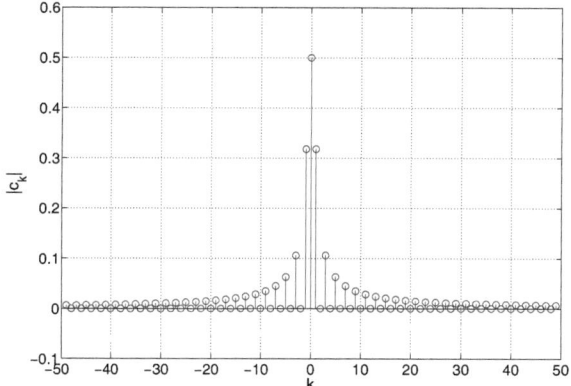

Figure 6.3: *Amplitude spectrum of the complex coefficients of the Fourier expansion of a periodic rectangular pulse function. The amplitude values are plotted against the running index k. The envelope follows a $\sin x/x$ curve.*

frequency $f_0 = 1/T$ accordingly to the index k, i.e. $f_k = k\,f_0 = k/T$. Therefore, the spacing is $1/T$. This is a key relationship in the analysis of discrete-time signals: it states that *increasing the observation time of the time signal improves the frequency resolution of its (discrete) Fourier transform* (see par. 6.2.3).

The considered example also points out the issue of the *truncation* of the Fourier series. In fig. 6.2 it can be observed that the approximation becomes better for an increasing number of considered terms. Nevertheless, the approximating function (i.e. the truncated series) show persisting oscillations which are larger at the edges of the rectangular pulse. The series expansion of a *sectionally* differentiable function[5] converges uniformly to the function in any subinterval excluding the point of discontinuity. At such points, if a finite number of terms is used for the expansion, the error in the approximation remains constant, whatever the number of the considered terms is. The approximating function (i.e. the summation of the harmonic terms) shows a typical oscillating behaviour in the vicinity of the discontinuities, which is called the *Gibbs' phenomenon*, from the name of the first scientist who investigated

---

[5] function whose domain is divided into an arbitrary number of sub-intervals within which the function is differentiable.

the problematic at the end of the eighteenth century. It can be shown that, for a fixed value of $t \neq 0$, the error tends to zero for $n \to \infty$, where n is the number of considered harmonics ([4]). Nevertheless, as n increases the error remains essentially constant; increasing n has no effect on the size of the over-swing: it merely compresses the phenomenon into a decreasing range of t near the discontinuities. This latter effect can be appreciated by comparing figs. 6.4 and 6.5, which show a blow-up of the rising edge of the pulse for series of twenty terms and one-hundred terms, respectively.

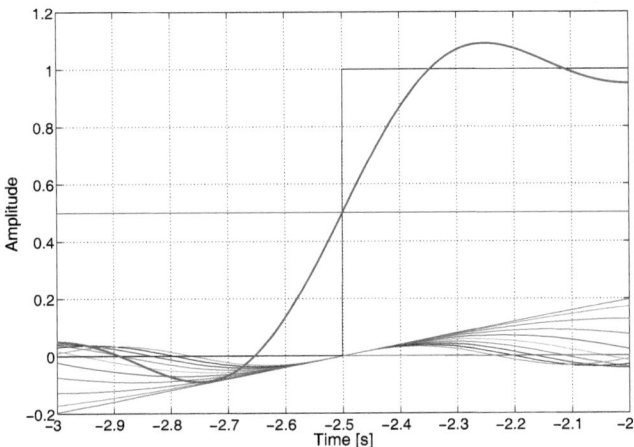

Figure 6.4: *Gibbs' phenomenon: oscillations in the vicinity of one point of discontinuity (20 terms considered).*

6.1.3 The Fourier Transform

As mentioned, increasing the period T leads to a closer spacing $\Delta f = 1/T$ of the Fourier coefficients c_k. Extending the period to infinity would lead to a non-periodic function $s(t)$ and to an infinitely close spacing between its frequency components:

$$\lim_{T \to \infty} \Delta f = 0 \tag{6.16}$$

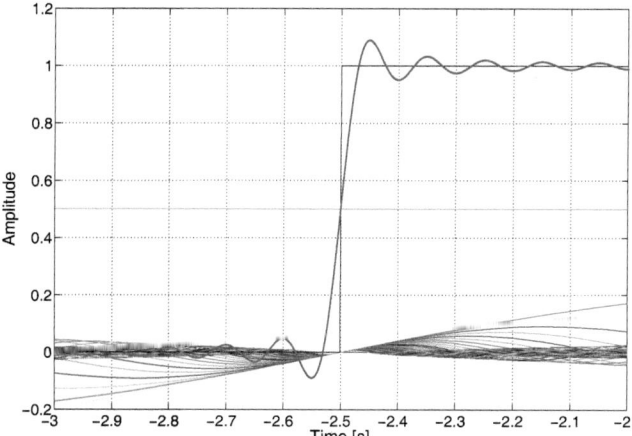

Figure 6.5: *Gibbs' phenomenon: oscillations in the vicinity of one point of discontinuity (100 terms considered). Comparison with fig. 6.4 shows that increasing the number of harmonics does not reduce the amplitude of the oscillations but compresses the ringing closer to the discontinuity point.*

As a consequence, the discrete index k would be substituted by the continuous variable f, so that the discrete Fourier coefficients of eq.(6.14) are substituted by a continuous function of f:

$$S(f) = \int_{-\infty}^{\infty} s(t) \, e^{-j2\pi ft} \, dt \,, \qquad (6.17)$$

and the sum in eq.(6.13) is substituted by the integral:

$$s(t) = \int_{-\infty}^{\infty} S(f) \, e^{j2\pi ft} \, df. \qquad (6.18)$$

The eqs. (6.17) and (6.18) are called, respectively, the *Fourier transform* and the *inverse Fourier transform*.

Substituting eq.(6.17) in eq.(6.18) leads to the Fourier's integral theorem ([9]):

$$s(t) = \int_{-\infty}^{\infty} \left[\int_{-\infty}^{\infty} s(t) e^{-j2\pi ft} dt \right] e^{j2\pi ft} df. \qquad (6.19)$$

The conditions for eq. (6.19) to hold, that is, the conditions for $s(t)$ to be Fourier *transformable* are:

- the integral of $|s(t)|$ from $-\infty$ to ∞ exists;
- any discontinuities in $s(t)$ are finite;
- $s(t)$ has a finite number of maxima and minima in any finite interval[6].

It follows that periodic signals do not have a Fourier transform in strict sense [7]. Nevertheless it is possible to consider the Fourier transform of their Fourier series expansion. Given a periodic function $s_p(t)$ and defining the basic function of $s_p(t)$ as $s(t)$, the Fourier transform of $s(t)$, calculated in kf_0 is:

$$S(kf_0) = S\left(\frac{k}{T}\right) = \int_{-\infty}^{\infty} s(t) e^{-j2\pi \frac{k}{T} t} dt = \int_{-T/2}^{T/2} s(t) e^{-j2\pi \frac{k}{T} t} dt, \qquad (6.20)$$

Confronting eq.(6.14) with eq.(6.20) leads to:

$$c_k = S\left(\frac{k}{T}\right) \qquad (6.21)$$

Eq. (6.21) confirms the observation made about the amplitude spectrum of the Fourier coefficients (see text in par. 6.1.2 and fig. 6.3).

From eqs. (6.13) and (6.21) follows:

$$s_p(t) = \sum_{k=-\infty}^{\infty} c_k e^{j2\pi \frac{k}{T} t} = \frac{1}{T} \sum_{k=-\infty}^{\infty} S\left(\frac{k}{T}\right) e^{j2\pi \frac{k}{T} t} \qquad (6.22)$$

---

[6]This last condition is not so stringent: some function with an infinite number of maxima and minima in a finite interval have a Fourier transform ([9]).
[7]Some not Fourier transformable functions might admit Fourier transform in *limit* sense. ([9], [4])

The transform of $s_p(t)$ is:

$$S_p(f) = \mathscr{F}\{s_p(t)\} = \mathscr{F}\left\{\frac{1}{T}\sum_{k=-\infty}^{\infty} S\left(\frac{k}{T}\right) e^{j2\pi\frac{k}{T}t}\right\} =$$

$$= \int_{-\infty}^{\infty}\left[\frac{1}{T}\sum_{k=-\infty}^{\infty} S\left(\frac{k}{T}\right) e^{j2\pi\frac{k}{T}t}\right] e^{-j2\pi ft} \, dt =$$

$$= \frac{1}{T}\sum_{k=-\infty}^{\infty} \int_{-\infty}^{\infty} S\left(\frac{k}{T}\right) e^{-j2\pi(f-\frac{k}{T})t} \, dt =$$

$$= \frac{1}{T}\sum_{k=-\infty}^{\infty} S\left(\frac{k}{T}\right) \delta\left(f - \frac{k}{T}\right), \qquad (6.23)$$

where $\delta(f)$ represents the *Dirac pulse*: it has zero amplitude for each value of the abscissa except at zero, where the amplitude is infinite; its integral from $-\infty$ to ∞ yields a unitary value.

By confronting eqs.(6.23) and (6.17) it can be seen that the spectrum of the periodic function $s_p(t)$ corresponds to the sampled spectrum of the basic function (one period), with spacing between samples dictated by the period T. This is a general remark in the signal theory: *periodic signals have a sampled spectrum and sampled signals have a periodic spectrum* (see par. 6.2.1).

Another interesting property of the Fourier transform is that real signals have a symmetric spectrum: the amplitude spectrum is *even*, that is, $A(f) = A(-f)$, whereas the phase spectrum is *odd*, that is, $-\phi(f) = \phi(-f)$. The symmetry applies also for the real- and imaginary parts of the spectrum, which are, respectively, even and odd. As an example, fig. 6.6 shows the amplitude of the Fourier transform of the same rectangular pulse function whose Fourier series expansion was calculated in par. 6.1.2. The Fourier pair is defined as:

$$A\, rect_\tau(t) \leftrightarrow A \cdot \tau \cdot sinc(\pi f \tau) = \frac{A\, sin(\pi f \tau)}{\pi f} \qquad (6.24)$$

In this case, since the time function is not only real, but also even-symmetric, the Fourier transform is real (see par. 6.1.2).

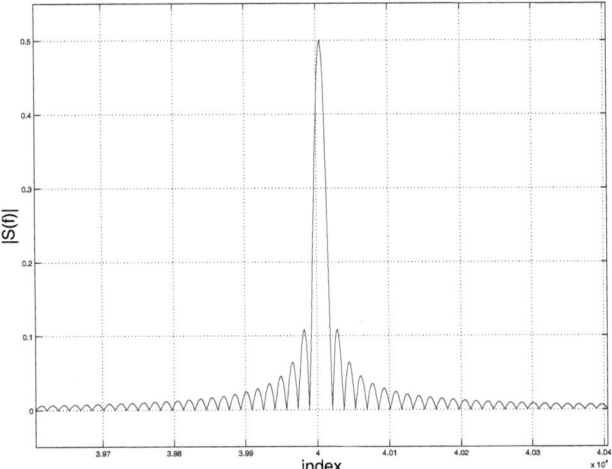

Figure 6.6: *The function $A\tau\,\text{sinc}(\pi f\tau)$ is the Fourier transform of the rectangular pulse $A\,\text{rect}_\tau(t)$, which was considered for the series expansion in par. 6.1.2, showing that real signals have symmetric Fourier transform (see text).*

Algebraic Space of the Signals

The exponential functions of the kind $e^{j2\pi\frac{k}{T}t}$ constitute an orthonormal base upon which the set of the time-signals has a vector space structure.

By setting $e^{j2\pi\frac{k}{T}t} = v_k$ and defining the inner product as:

$$<v_i, v_j^*> = \int_{-T/2}^{T/2} v_i(t)\,v_j^*(t)\,dt = \int_{-T/2}^{T/2} e^{j2\pi\frac{i}{T}t}\,e^{-j2\pi\frac{j}{T}t}dt =$$
$$= \begin{cases} 1 & \text{for } i \neq j \\ 0 & \text{for } i = j \end{cases} \tag{6.25}$$

each signal $s(t)$ can be expressed as *linear combination* of the vector of the base by means of opportune coefficients c_k:

$$s = \sum_{k=-\infty}^{\infty} c_k v_k \qquad (6.26)$$

The coefficients c_k are therefore calculated as the *projection* of vector s onto the basis v_k:

$$\begin{aligned}
c_k &= \frac{<s(t), v_k^*>}{||v_k||} = \frac{<s(t), v_k^*>}{<v_k, v_k>} = \\
&= \frac{\int_{-T/2}^{T/2} s(t) \left[e^{j2\pi \frac{k}{T}t}\right]^* dt}{\int_{-T/2}^{T/2} e^{j2\pi \frac{k}{T}t} \left[e^{j2\pi \frac{k}{T}t}\right]^* dt} = \frac{\int_{-T/2}^{T/2} s(t) e^{-j2\pi \frac{k}{T}t} dt}{\int_{-T/2}^{T/2} e^{j2\pi \frac{k}{T}t} e^{-j2\pi \frac{k}{T}t} dt} = \\
&= \frac{1}{T} \int_{-T/2}^{T/2} s(t) e^{-j2\pi \frac{k}{T}t} dt \quad , \qquad (6.27)
\end{aligned}$$

which coincide with the expressions of the Fourier coefficients given in eq. (6.14).

By making use of the introduced formalism, the inner product of two finite-energy signals $x(t)$, and $y(t)$ is given by:

$$\begin{aligned}
<x(t), y(t)> &= \int_{-\infty}^{\infty} x(t) y^*(t) \, dt = \\
&= \int_{-\infty}^{\infty} y^*(t) \left[\int_{-\infty}^{\infty} X(f) e^{j2\pi ft} \, df\right] dt = \\
&= \int_{-\infty}^{\infty} X(f) \int_{-\infty}^{\infty} y^*(t) e^{j2\pi ft} \, df \, dt = \\
&= \int_{-\infty}^{\infty} X(f) \left[\int_{-\infty}^{\infty} y(t) e^{-j2\pi ft} \, dt\right]^* df = \\
&= \int_{-\infty}^{\infty} X(f) Y^*(f) \, df \qquad (6.28)
\end{aligned}$$

The expression of of the inner product of a signal with itself corresponds to the

Parseval relationship for finite-energy signals (see eq.(6.15))

$$< x(t), x(t) >= ||x(t)||^2 = \int_{-\infty}^{\infty} |x(t)|^2\, dt = \int_{-\infty}^{\infty} |X(f)|^2\, df = \mathscr{E}, \qquad (6.29)$$

where $\mathscr{E}$ is the *energy* of $x(t)$ (see par.6.1.4).

From the previous relationship follows that the integrand must represent the frequency distribution of the signal energy, or the *energy density spectrum* $\mathscr{S}(f)$:

$$X(f) \cdot X^*(f) = |X(f)|^2 = \mathscr{S}(f) \qquad (6.30)$$

6.1.4 Cross- and Autocorrelation; the Wiener-Kintchine Theorem

Finite-Energy Signals

Given two signals $x(t)$, and $y(t)$, the degree of similarity they have can be calculated by means of the *cross-correlation*. For finite-energy signals, the cross-correlation is defined as:

$$R_{xy}(\tau) = \int_{-\infty}^{\infty} x(t) y^*(t+\tau)\, dt\,; \qquad (6.31)$$

whereas, the *autocorrelation* is defined as the cross-correlation of a signal with itself. It gives a measure of the similarity between observations as a function of the time separation between them:

$$R_{xx}(\tau) = R_x(\tau) = \int_{-\infty}^{\infty} x(t) x^*(t+\tau)\, dt \qquad (6.32)$$

The computation of the Fourier transform of the autocorrelation leads to:

$$\mathcal{F}\{R_x(\tau)\} = \int_{-\infty}^{\infty} \left[\int_{-\infty}^{\infty} x(t) x^*(t+\tau) \, dt \right] e^{-j2\pi f \tau} \, d\tau =$$
$$= X(f)^* \cdot X(f) = |X(f)|^2 = \mathscr{S}(f) \quad (6.33)$$

The relationship (6.33) is known as the *Wiener-Kintchine Theorem* for finite-energy signals: the Fourier transform of the autocorrelation $R_x(\tau)$ of a signal $x(t)$ equals the *energy density spectrum* $\mathscr{S}(f)$ of the signal. $\mathscr{S}(f)$ expresses the energy content of the signal $x(t)$ at each given frequency f.[8]

Integration eq.(6.33) over the frequency axis gives the total energy of the signal: the same results of the Parseval relationships in eq.(6.29). It is interesting to notice that *the signal autocorrelation computed at the origin ($\tau = 0$) coincides with the total signal energy*:

$$R_x(\tau)|_{\tau=0} = \mathcal{F}^{-1}\{X(f)^* \cdot X(f)\}|_{\tau=0} = \int_{-\infty}^{\infty} X(f)^* \cdot X(f) \, e^{j2\pi f \tau} \, df \bigg|_{\tau=0} =$$
$$= \int_{-\infty}^{\infty} X(f)^* \cdot X(f) \, df = \int_{-\infty}^{\infty} \mathscr{S}(f) \, df = \mathscr{E} \quad (6.34)$$

Periodic Signals

Finite-power signals are not square integrable: in this case, the definition of autocorrelation must be adapted. Given a periodic signal $x(t)$ of period T, its autocorrelation is defined as:

$$R_x(\tau) = \frac{1}{T} \int_{-T/2}^{T/2} x(t) x^*(t+\tau) \, dt \quad (6.35)$$

Writing $x(t+\tau)$ as its Fourier expansion leads to:

---

[8]Taking signals in [V] over a unity-resistor, the autocorrelation has the dimension $[\frac{V^2}{\Omega} \cdot s]$. Integrating over time gives the energy spectrum the dimension $[\frac{V^2}{\Omega} \cdot s \cdot s] = [W \cdot s \cdot \frac{1}{Hz}] = [\frac{J}{Hz}]$, that is, a spectral energy density.

$$R_x(\tau) = \frac{1}{T}\int_{-T/2}^{T/2} x(t)\left[\sum_{k=-\infty}^{\infty} c_k\, e^{j2\pi\frac{k}{T}(t+\tau)}\right]^* dt =$$

$$= \frac{1}{T}\sum_{k=-\infty}^{\infty} c_k^*\, e^{-j2\pi\frac{k}{T}\tau} \int_{-T/2}^{T/2} x(t)\, e^{-j2\pi\frac{k}{T}t}\, dt =$$

$$= \sum_{k=-\infty}^{\infty} c_k c_k^*\, e^{-j2\pi\frac{k}{T}\tau} = \sum_{k=-\infty}^{\infty} |c_k|^2\, e^{-j2\pi\frac{k}{T}\tau} \qquad (6.36)$$

The computation of the Fourier transform of the autocorrelation leads to:

$$\mathscr{F}\{R_x(\tau)\} = \int_{-\infty}^{\infty}\left[\sum_{k=-\infty}^{\infty} |c_k|^2\, e^{-j2\pi\frac{k}{T}\tau}\right] e^{-j2\pi f\tau}\, d\tau$$

$$= \sum_{k=-\infty}^{\infty} |c_k|^2\, \delta(f + \frac{k}{T}) = \mathscr{S}(f) \qquad (6.37)$$

The relationship (6.37) is known as the *Wiener-Kintchine Theorem* for periodic signals: the quantity $\mathscr{S}(f)$ is called *power density spectrum*. It expresses the power of the various harmonics forming the periodic signal $x(t)$.[9]

Analogously to (6.34), the signal autocorrelation computed at the origin ($\tau = 0$) equals the average signal power over a period:

$$R_x(\tau)|_{\tau=0} = \int_{-\infty}^{\infty} \mathscr{S}(f)\, df = \mathscr{P} \qquad (6.38)$$

6.1.5 Linear Time Invariant Systems

In this context, the concept of "system" applies to an electric network where two terminals are defined as "input" and other two are defined as "output". The system transforms an input signal $x(t)$ in a output signal $y(t)$. Indicating the transformation by "T", the process can be described as: $y(t) = T[x(t)]$. If the system is "time invariant", the operator "T" does not depend on time. If the property of "linearity"

---

[9] Taking signals in [V] over a unity-resistor, the autocorrelation has the dimension $[\frac{V^2}{\Omega}]$. Integrating over time gives the power spectrum the dimension $[\frac{V^2}{\Omega}\cdot s] = [\frac{W}{Hz}]$, that is, a spectral power density.

applies, given two signals $x_1(t)$, and $x_2(t)$, which form the input $\alpha x_1(t) + \beta x_2(t)$, with α, β two numerical constant, for the output $y(t)$ the relationship is valid:

$$y(t) = T[\alpha x_1(t) + \beta x_2(t)] = \alpha T[x_1(t)] + \beta T[x_2(t)] \tag{6.39}$$

If both properties apply, the system is said LTI. As mentioned in par.6.1.2, as harmonic inputs are eigenfunctions of LTI systems, they are modeled by their *frequency response*, $H(f)$, defined as the ratio between the Fourier transform of the output signal and the Fourier transform of the input signal:

$$H(f) = \frac{Y(f)}{X(f)}, \tag{6.40}$$

where $X(f) = \mathscr{F}\{x(t)\}$, and $Y(f) = \mathscr{F}\{y(t)\}$.

The function $H(f)$ is called the system function and depends on the terminals which are respectively chosen as input and output. The system function of a LTI system is represented by a complex function and can be characterized by its amplitude $A(f)$ and phase $\varphi(f)$:

$$H(f) = A(f) e^{j\varphi(f)}, \tag{6.41}$$

From eq.(6.40) it is possible to see that the spectrum of the output signal can be calculated by multiplying the spectrum of the input signal by the frequency response of the system. In order to analyze the behaviour of the network in the time domain, the inverse Fourier transforms of the involved functions are considered, so that ($\leftrightarrow$ indicates a Fourier pair):

$$Y(f) = X(f) \cdot H(f) \leftrightarrow y(t) = x(t) * h(t) = \int_{-\infty}^{\infty} x(\tau) h(t-\tau) \, d\tau \tag{6.42}$$

The operator "$*$" is the *convolution* operator. Eq.(6.42) shows a fundamental property of the Fourier representation: *product in the frequency domain corresponds to convolution in the time domain (and vice versa)*.

The inverse transform of the system function, which is the function $h(t)$, is called the *impulse response* of the system. It can be characterized directly in the time domain as the system output caused by a Dirac pulse put at input. The right-hand side equation of the relationship (6.42) can be derived directly in the time domain by some considerations based on the properties of the Dirac pulse. First of all the *sampling* property of the Dirac pulse:

$$x(t)\delta(t-t_0) = x(t_0), \tag{6.43}$$

which leads to the consideration that each signal can be thought of as formed by an infinite sum of Dirac pulses:

$$\begin{aligned} x(t) &= x(t_1)\delta(t-t_1) + x(t_2)\delta(t-t_1) + \ldots \\ &= \int_{-\infty}^{\infty} x(\tau)\delta(t-\tau)\,d\tau = x(t) * \delta(t) \end{aligned} \tag{6.44}$$

that is, the property of the Dirac impulse to be the neutral element of the convolution. Due to the linearity of the system (see eq.(6.39)), the output corresponding to the excitation $x(t)$ can be calculated as the sum of the single outputs caused by the single inputs in eq.(6.44):

$$\begin{aligned} y(t) &= x(t_1)h(t-t_1) + x(t_2)h(t-t_2)) + \ldots = \\ &= \int_{-\infty}^{\infty} x(\tau)h(t-\tau)\,d\tau = x(t) * h(t) \end{aligned} \tag{6.45}$$

that is: the response of a LTI system to a given excitation can be calculated as the convolution of the input signal with the impulse response of the system.

6.1.6 Modulation

Multiplying a signal with a sinusoidal function of a given frequency corresponds to shifting its spectrum around the frequency of the sinusoidal term. This operation is called *modulation* and it is exploited for signal transmission: base-band signals are moved to a region of the frequency axis which is suitable for transmission (modulation), and moved bach to base-band (zero-frequency) upon reception (demodulation). As in the relationship (6.42), this property is based on the duality between the time- and the frequency domain: the product in one domain corresponds to the convolution in the other domain. Since:

$$\mathscr{F}\{cos(2\pi f_0 t)\} = \frac{\delta(f-f_0) - \delta(f+f_0)}{2}, \tag{6.46}$$

for the modulation is:

$$y(t) = s(t) \cdot cos(2\pi f_c t) \leftrightarrow Y(f) = S(f) * \left[\frac{\delta(f-f_c) - \delta(f+f_c)}{2}\right], \tag{6.47}$$

where the subscript "c" indicates the *carrier* frequency. Eq. (A) show that the frequency content of the base band signal $s(t)$ has been moved around the carrier frequency (also in the negative part of the spectrum).

6.1.7 The analytic signal and the Hilbert Transform

As representative of physical processes, signals are described by real functions. Nevertheless, they can be advantageously represented in complex notation. The complex signal $s_a(t)$, associated to the real signal, is called the *analytic signal*:

$$s_a(t) = A(t)\, e^{j\phi(t)}\, e^{j\omega_c(t)} \tag{6.48}$$

The following equation describes the relationships between the two signals (see also par. 5.3):

$$\begin{aligned} s(t) &= \mathfrak{Re}\{s_a(t)\} = \mathfrak{Re}\{A(t)\, e^{j\phi(t)}\, e^{j\omega_c(t)}\} = \\ &= \mathfrak{Re}\{A(t)\, cos[\omega_c(t) + \phi(t)] + j\, A(t)\, sin[\omega_c(t) + \phi(t)]\} = \\ &= A(t)\, cos[\omega_c(t) + \phi(t)] \end{aligned} \tag{6.49}$$

where $s(t)$ is the real signal, ω_c the carrier angular frequency, and $A(t)$ and $\phi(t)$

respectively the amplitude and phase of the base-band signal, i.e. $A(t)\cos[\phi(t)]$.

Writing signals in the complex notation implies that only the contribution of the positive (or negative) frequencies shows up in spectrum. In order to demonstrate this, we consider the positive half of the spectrum, which we define $S^+(f)$. This portion of the spectrum is given by the product of the whole spectrum $S(f)$ by the function *unit step* considered in the frequency domain:

$$S^+(f) = S(f) \cdot U(f) , \qquad (6.50)$$

where

$$U(f) = \begin{cases} 1 & (f > 0) \\ 1/2 & (f = 0) \\ 0 & (f < 0) \end{cases} \qquad (6.51)$$

The product in the frequency domain corresponds to a convolution in the time domain (see par. 6.1.5), therefore, from eq. (6.50) follows:

$$\mathscr{F}^{-1}\{S^+(f)\} = s^+(t) = s(t) * \mathscr{F}^{-1}\{U(f)\} = s(t) * \left[\frac{1}{2}\delta(t) - \frac{1}{j2\pi t}\right] =$$
$$= \frac{1}{2}s(t) + \frac{j}{2}s(t) * \frac{1}{\pi t} \qquad (6.52)$$

The second term on the right-hand side of eq. (6.52) (excluding the factor j) is called the *Hilbert transform* of the signal $s(t)$. Making use of the notation $\hat{s}(t)$ for the Hilbert transform, we have the relationship:

$$\hat{s}(t) = s(t) * \frac{1}{\pi t} = \int_{-\infty}^{\infty} \frac{1}{\pi} \cdot s(\tau) \cdot \frac{1}{t-\tau} d\tau \qquad (6.53)$$

Therefore:

$$s^+(t) = \frac{1}{2}[s(t) + j\hat{s}(t)] , \qquad (6.54)$$

where the apex "+" denotes the fact that the spectrum of such a signal has only positive frequency components. It is possible to show that the analytic signal introduced by the relationship (6.48) corresponds to the (half) sum of the original signal and its Hilbert transform, that is, the complex polar notation of eq. (6.48) can be expressed in Cartesian form by means of the real signal and its Hilbert transform:

$$s_a(t) = A(t)\, e^{j\phi(t)}\, e^{j\omega_c(t)} = s^+(t) = \frac{1}{2}[s(t) + j\hat{s}(t)], \qquad (6.55)$$

The effect of the Hilbert transform is to rotate the signal phase by an amount of $\pi/2$, letting the amplitude unaltered. As a consequence, the sequence of the *quadrature* samples represents the Hilbert transform of the sequence of the *in-phase* samples (see par. 5.3). Adding the signal and its "turned" version together in a unique complex signal let one of two sidebands disappear.[10] This is of great advantage in the processing of the occultation *open loop* data: since the carrier signal is moving along the spectral time sequence, in a symmetric representation of the spectrum it would be difficult to distinguish if the peak has passed the zero-frequency.

This argumentation is exemplified by figs. 6.7 and 6.8 which show two power density spectra computed from the data of the VEX-VeRa occultation pass on DoY220-2006, orbit #109 (power values are not calibrated). The spectra are derived respectively from the *in-phase* sequence $I(t)$ and from the complex sequence $I(t) + jQ(t)$ (see par. 5.3), for the same time interval (1 s observation time). Observation and comparison of the two spectra point out important characteristics. First of all, the spectrum of the real signal ($I(t)$) is symmetric. Furthermore, the signal peak of the power spectrum computed from the complex signal is $6\,dB$ larger than each peak of the power spectrum of the real signal, as it was to expect, since the amplitude correction factor $1/2$ was not considered in the complex sum (see eq. (6.54)). It has to be noticed that summing both channels, or signal components, in a unique complex signal does not cancel the background noise in one half of the spectrum. This can be explained by considering that the sinusoidal signal of the receiver oscillator has a *coherent* phase relationship with the incoming signal, since the phase of the incoming signal is *deterministic*. This implies that the phase difference of $\pi/2$

---

[10] This result is also in agreement with the property of symmetry of the spectrum of real signals (see 6.1.3): as soon as the signal becomes complex the symmetry in the spectrum is lost.

between both channels is maintained at each moment. This is not the case for the superimposed thermal noise: since its phase belongs to a *stochastic* process with *uniform distribution* between $\pm\pi$, there is no coherency between the phase values of the noise on the two channels. Therefore it does not cancel in the complex sum as the signal does. Furthermore, the level of the noise floor shown in the spectrum of the complex signal is roughly $3\,dB$ higher than in the spectrum of the in-phase sequence, as it was to expect due to the summation of the independent noise processes contained in the two channels. (A description of the statistical properties of white noise is given in Ch. 8)

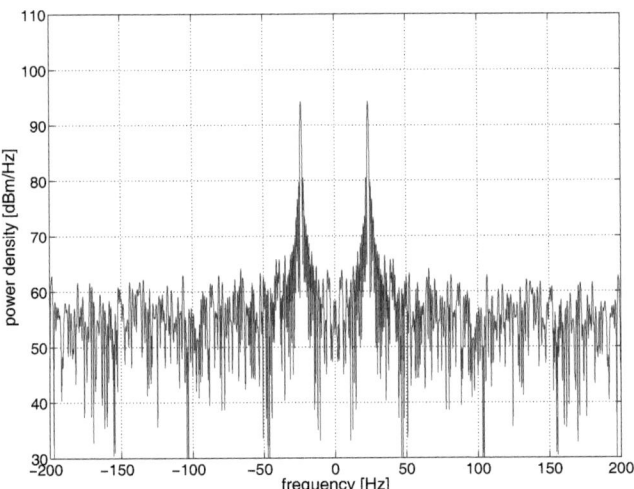

Figure 6.7: *Spectral power density of 1 s data from the in-phase channel ($I(t)$). Since the signal is real, the spectrum is symmetric. Both positive- and negative frequency contribution to the signal are present. Data from the VEX-VeRa occultation pass on DoY220-2006, orbit #109 (power values not calibrated).*

The base-band component of the analytic signal is:

$$\underline{s}(t) = A(t)\, e^{j\phi(t)}, \tag{6.56}$$

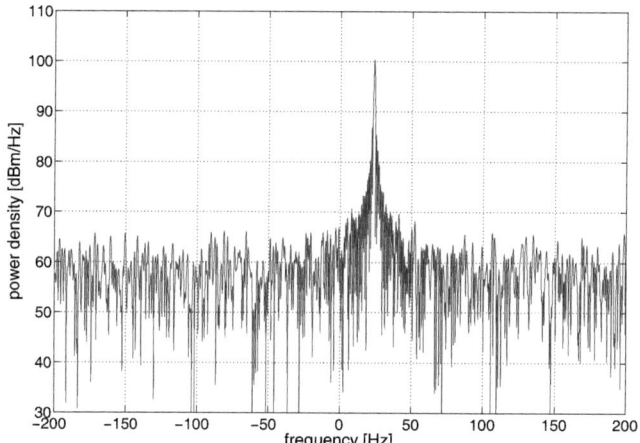

Figure 6.8: *Spectral power density of 1 s complex time signal. Since the signal is complex, the spectrum is no longer symmetric. Only the positive frequency contribution to the signal is present, whereas the noise shows both contributions (see text for comments). Comparison with fig. 6.7 points out a difference of 6 dB in the signal level, as it was to expect, since the factor 1/2 was not considered in the complex sum (see eq. (6.54)). The level of the noise floor is roughly 3 dB higher than in the spectrum of the in-phase sequence, as it was to expect due to the summation of the independent noise processes contained in the two channels. Data from the VEX-VeRa occultation pass on DoY220-2006, orbit #109 (power values not calibrated).*

which in Cartesian form is:

$$\underline{s}(t) = a(t) + jb(t), \tag{6.57}$$

where the relationships hold:

$$\begin{cases} A(t) = \sqrt{a^2(t) + b^2(t)} \\ \phi(t) = \arctan\left[\frac{b(t)}{a(t)}\right] \end{cases} \tag{6.58}$$

The real- and imaginary parts of $\underline{s}(t)$, i.e. $a(t)$ and $b(t)$, are called the "base-band analogue components" of the RF signal. They are also known as the "in-phase", and the "quadrature" components, respectively, as they differ by a constant phase off-set of 90°.

6.2 Discret-Time Signals[11]

A discrete-time signal is a sequence of values derived by looking at the value of an analog signal only at given time instants. The obtained values are arranged in a numeric sequence characterized by a running index included in brackets: being $s(t)$ the originating analog signal, the extracted sequence is $s[n]$. Discrete-time sequences assume continuous values, so that their imagine is a subset of $\mathbb{R}$, or $\mathbb{C}$. It is possible to discretize the imagine of the sequence by splitting it up into a certain numbers of intervals and assigning the same value to all points of the sequence which fall in the same interval. The obtained sequence is then called "digital signal".

The basic relationships for discrete-time (or digital) signals derive from the analogue relationships for continuous-time signals, so that, in principle, they can be obtained by substituting time-dependency with indexes, and integrals with summation. Therefore, concepts which were already presented for analog signal and which can be directly derived from these will be omitted. Only basic differences between analog signals and sequences, along with major specific topics of the discrete-time signal theory, will be shortly addressed.

6.2.1 Sampling theorem

A discrete-time signal, or sequence $s[n]$ can be obtained from an analog signal $s(t)$ through a process called *sampling*. In this process the original signal is *sampled* by means of an infinite series of Dirac pulses (Dirac pulse train):

$$s_s(t) = s(t) \cdot \sum_{n=-\infty}^{\infty} \delta(t - nT_s), \qquad (6.59)$$

where $s_s(t)$ is the sampled signal, T_s is the *sampling period* or *sampling time*,

---

[11] An extensive treatment of this subject can be found in [71].

and $\delta(t)$ is the Dirac pulse or Dirac delta function.[12] Since the Fourier transform of an infinite series of Dirac pulses is an infinite series of Dirac pulses as well, the Fourier transform of the sampled signal in eq.(6.59) is:

$$S_s(f) = S(f) * \left[\frac{1}{T_s} \sum_{n=-\infty}^{\infty} \delta\left(f - \frac{n}{T_s}\right) \right] \qquad (6.60)$$

The reciprocal of the sampling period is the *sampling frequency*, or *sampling rate* $(1/T_s = f_s)$. Eq.(6.60) establishes that the Fourier transform of a sampled signal consist of an infinite number of amplitude-scaled replica of the transform of the original signal, centered around multiples of the sampling frequency. Thus, in order to reconstruct the original signal from the spectrum of the sampled signal, it is necessary to isolate the first replica (around zero-frequency) by means of an opportune *filter* and apply the inverse Fourier transform. This can only be done if the sampling frequency is at least as large as twice the band occupation of the signal, otherwise replicas would overlap altering the frequency content of the signal, giving rise to a phenomenon called *aliasing*. The minimum limit for the sampling frequency is set by the *Nyquist criterion*:

$$f_s \leq 2\,BW, \qquad (6.61)$$

where BW is the (monolateral) bandwidth of the signal to be sampled. Therefore, in order to confine a signal in the bandwidth BW, a pre-sampling filtering has to be performed.

The original signal can be thought of as the inverse Fourier transform of the product of the frequency response of the reconstructing filter, $F(f)$, with the spectral

---

[12] It has to be noticed that, formally, $s_s(t)$ is still a continuous-time signal. The relevant discrete-time sequence can be obtained from $s_s(t)$ by taking all non-zero values and ordering them by a discrete index.

replicas:[13]

$$s(t) = \mathscr{F}^{-1}\left\{F(f) \cdot \sum_{n=-\infty}^{\infty} \frac{1}{T_s} S\left(f - \frac{n}{T_s}\right)\right\}, \qquad (6.62)$$

that is, as the convolution between the sampled signal and the impulse response of the filter, $f(t)$:

$$s(t) = s_s(t) * f(t) \qquad (6.63)$$

Considering the frequency response of the ideal filter (rectangular filter), whose phase is identically null and whose amplitude is defined as:

$$|F(f)| = \begin{cases} T_s & for |f| \leq f_s/2 \\ 0 & for |f| > f_s/2 \end{cases} \qquad (6.64)$$

its impulse response is:

$$f(t) = T_s\, f_s\, sinc(\pi f_s t) = sinc(\pi f_s t)\ , \qquad (6.65)$$

where $sinc(x) = sin(x)/x$. Substituting eq.(6.59) and eq.(6.65) in eq.(6.63) yields:

$$s(t) = \int_{-\infty}^{\infty} s(\tau) \sum_{n=-\infty}^{\infty} \delta(\tau - nT_s) \cdot sinc[\pi f_s(t-\tau)]\, d\tau =$$
$$= \sum_{n=-\infty}^{\infty} s(nT_s)\, sinc[\pi f_s(t - nT_s)] \qquad (6.66)$$

The relationship (6.66) is known as the "interpolation formula": it shows how to reconstruct the original analog signal from its samples. Since the ideal filter is

---

[13]The factor which should compensate for the scaling of the replicas (see eq. (6.60) is considered embedded in the filter $F(f)$, which, therefore, has constant amplitude T_s over the bandwidth BW.

not physically realizable (infinite slope of the edge), the sampling frequency has to be set to higher values than the Nyquist criterion for a filter with finite slope not to cause aliasing.

6.2.2 The Fourier Transform of discrete-time signals

Since the index n is a dimensionless integer, important differences arise between continuous-time sinusoidal (or complex exponential) signals and the sequences which derive from them.

Given a sequence $A\,cos(\omega_0 n + \varphi)$, because of the index n, which is a dimensionless integer, the (angular) frequency ω_0 has no longer the dimension $[s^{-1}]$, as in the case of analog signals: it must be given in radians, for the argument of the trigonometric function to be a phase. Therefore, the sequence $A\,cos(\omega_0 n + \varphi)$ coincide with the sequence $A\,cos[(\omega_0 + 2\pi r)n + \varphi]$, $\forall r \in \mathbb{Z}$.

This implies that frequency spectra have a 2π periodicity:[14]

$$X(\omega) = X(e^{j\omega}) = X(e^{j(\omega+2\pi)}), \qquad (6.67)$$

In the spectrum of the sequence, the "frequencies" $\pm \pi$ correspond to the frequencies $\pm \frac{f_s}{2}$ of the spectrum of the sampled signal. Therefore, the Fourier transform of a sequence is completely defined by taking the interval $[-\pi, \pi]$. Given a sequence $x[n]$, the Fourier pair of transform and inverse transform is:

$$X(e^{j\omega}) = \sum_{n=-\infty}^{\infty} x[n]\,e^{-j\omega n} \qquad (6.68)$$

$$x[n] = \frac{1}{2\pi} \int_{-\pi}^{\pi} X(e^{j\omega n})\,e^{j\omega n}\,d\omega \qquad (6.69)$$

An interesting observation can be made about the Fourier transform of a discrete sequence: $X(e^{j\omega})$ can be thought of as a continuous-domain periodic function hav-

---

[14]The periodicity of a sampled signal was also a consequence of the sampling theorem (see par.6.2.1).

ing the Fourier series expansion given by eq.(6.68), where the Fourier coefficients, calculated by means of (6.69), are the sequence itself.

Determining the class of sequences for which the relationship (6.69) holds is equivalent to considering the convergence of the sum in eq.(6.68), that is, the conditions that must be satisfied such that $|X(e^{j\omega})| < \infty$, $\forall \omega$. A sufficient condition for convergence can be found from following relationship:

$$|X(e^{j\omega})| = |\sum_{n=-\infty}^{\infty} x[n] e^{-j\omega n}|$$
$$\leq \sum_{n=-\infty}^{\infty} |x[n]| |e^{-j\omega n}| \leq \sum_{n=-\infty}^{\infty} |x[n]| < \infty \qquad (6.70)$$

Thus, if $x[n]$ is absolutely summable, then $X(e^{j\omega})$ exists ([71]).

6.2.3 The Discrete Fourier Transform (DFT)

A sequence $x[n]$ for which is:

$$x[n] = x[n + rN], \qquad \forall n, r \in \mathbb{Z}; \; \forall N \in \mathbb{N} \qquad (6.71)$$

is periodic. It can be expanded in Fourier series:

$$x[n] = \frac{1}{N} \sum_{k=0}^{N-1} X[k] e^{j\frac{2\pi}{N}kn}, \qquad (6.72)$$

where $X[k]$ are the Fourier coefficients. The basis for the expansion is given by the complex exponential sequences $e^{j\frac{2\pi}{N}kn}$, $k = 0, ..., N$, which are periodic of period N. For this reason, the sum in (6.72) is limited to N terms.

The Fourier coefficients $X[k]$ are obtained from $x[n]$ by the relationship:

$$X[k] = \sum_{n=0}^{N-1} x[n] e^{-j\frac{2\pi}{N}kn}, \qquad (6.73)$$

From eq.(6.73) derives that the sequence $X[k]$ is also periodic, with period N.

It was already mentioned that a discrete-time sequence corresponds to a sampled signal and therefore has periodic spectrum. From the DFT theory derives that, in addition to periodicity, the spectrum of a *periodic sequence* is also *discrete* (see par. 6.1.3).

The following relationships holds for sampled spectra obtained by the DFT:

$$N = f_s \cdot T = \frac{1}{\Delta t} \cdot T \qquad (6.74)$$

where f_s is the sampling frequency (or sample rate), $\Delta t = 1/f_s$ is the sampling time, and T is the length of the considered portion of the originating analog signal. As mentioned in par. 6.1.2, the spacing between samples in the frequency domain corresponds to $1/T$, which implies that the *frequency resolution improves as the observation time increases*.

Zero-Padding (ZP)

This method was adopted in the OL SW in order to enhance the performances of the spectral analysis, used to derive the mix-signals of the various processing stages (see par. 7.1.4).

Digital processing of sampled signals means dealing with periodic sequences. This is due to the fact that, since the samples are processed in sets of given length, a processing algorithm, i.e. the DFT, considers each set as one period of length N of a periodic sequence.[15] This implies an implicit multiplication, or *windowing*, of the analyzed samples with a rectangular unity function, which, in the frequency domain, corresponds to convolving the original spectrum by a $sinx/x$ curve (see eq. (6.42)).

When the DFT is performed on a given set of samples taken over a period T, the frequency resolution of the underlying continuous spectrum can be improved by means of *"Zero-Padding"* (ZP). This algorithm adds a series of zeros at the end of the sample set to be analyzed. The length of the new sample set corresponds to a time period $T' > T$, thus $1/N' < 1/N$ (see eq. (6.74)). It has to be stressed that no

---
[15]Given the sampling rate, the number of samples N depends on the observation duration T of the original analog signal.

new information is created: the ZP algorithm only allows the underlying continuous spectrum to be revealed in more detail.

Figs. 6.9 and 6.10 illustrate the difference between the DFT of a sinusoidal signal executed with- and without zero-padding, respectively, for two different values of the signal frequency. As mentioned above, the expected spectrum is of the form $sinx/x$, centered at the frequency of the sinusoidal signal. In the first case, the signal frequency was chosen to be an exact multiple of the observation period $T = 1\,s$, i.e. $f = 100\,Hz$, whereas in the second case it was $f = 99.78\,Hz$. In both cases, the red line shows the DFT without zero-padding, whereas the blue line corresponds to the DFT performed by means of the zero-padding algorithm, with an increasing factor in the number of the points $n = 2^6$.

It can be seen in fig.6.9 that all spectral component of the DFT without ZP are all null except for the one at the signal frequency, which presents the same amplitude of the "true" underlying spectrum (traced by means of the ZP algorithm).

In the other considered case, when the signal frequency is not an exact multiple of the observation period, the signal energy distributes among adjacent frequency bins, since the higher harmonics do not cancel in the sum (6.73) (fig.6.10). In this case, besides the component centered at the signal frequency, (whose amplitude is less than the true value), other spurious frequencies rise in the spectrum. This phenomenon is known as *leakage* ([2]). The effect of the zero-padding is to increase the *fundamental period* (i.e. the observation period) so that the gap between two adjacent harmonics of the Fourier expansion (base functions of the expansion, see par. 6.1.3) become smaller, therefore increasing the probability that the analyzed sine wave matches with one of the base functions.

The Fast Fourier Transform (FFT)

The amount of computations required for the DFT is approximately proportional to N^2. This number can be reduced by exploiting the properties of *complex conjugate symmetry* and *periodicity* of the complex exponentials forming the base of the DFT. A variety of algorithms have been developed in order to calculate the DFT at lower computational cost. They are collectively known as Fast-Fourier-Transform (FFT). By means of this kind of algorithms, the computational complexity is reduced to $O(N\,log_2 N)$ ([71]).

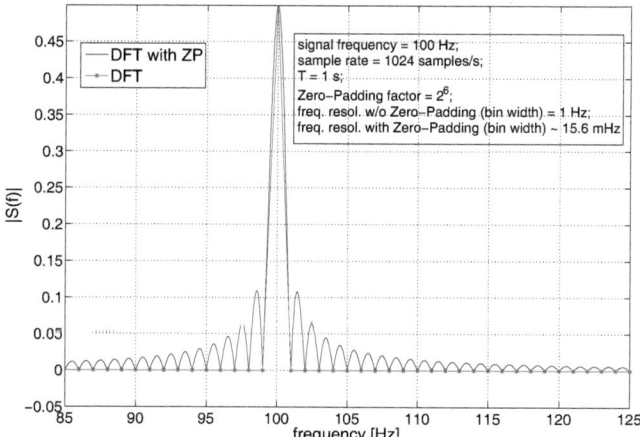

Figure 6.9: *Improvement of the spectral resolution by means of the Zero-Padding algorithm. The DFT of a sinusoidal signal of frequency $f = 100\,Hz$, observed over a period $T = 1\,s$, was calculated without ZP (red curve) and with ZP (blue curve). The frequency resolution (bin width) is increased from $1\,Hz$ to $15.6\,mHz$. Only the component centered at the signal frequency is to be seen in the spectrum obtained without ZP. The true spectrum remains "hidden".*

6.2.4 The "z-transform"

The z-transform for discrete-time signals plays the same role of the Laplace transform for continuous-time signals. Its a generalization of the Fourier transform, which allows a broader class of sequences to be analyzed, as the Fourier transform does not converge for all sequences.

The z-transform is defined by starting from the Fourier transform in eq.(6.68), repeated hereafter for convenience:

$$X(e^{j\omega}) = \sum_{n=-\infty}^{\infty} x[n]\, e^{-j\omega n} \;; \tag{6.75}$$

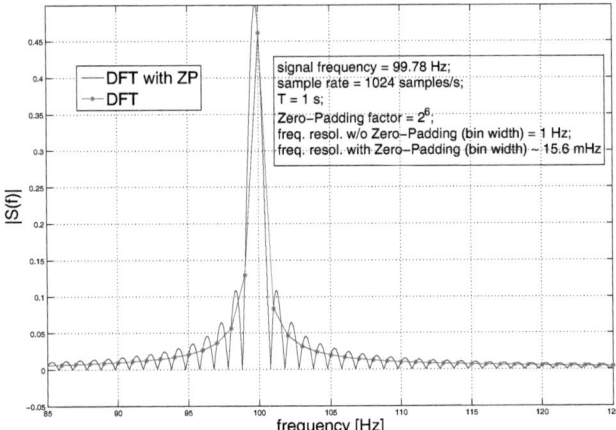

Figure 6.10: *Improvement of the spectral resolution by means of the Zero-Padding algorithm. The DFT of a sinusoidal signal of frequency $f = 99.78\,Hz$, observed over a period $T = 1\,s$, was calculated without ZP (red curve) and with ZP (blue curve). The frequency resolution (bin width) is increased from $1\,Hz$ to $15.6\,mHz$. In this case, the signal energy distributes among adjacent frequency bins, so that besides the component centered at the signal frequency (whose amplitude is less than the real value), other spectral components show up. This phenomenon is known as "leakage" ([2]).*

by letting $e^{-j\omega n} = z^{-n}$:

$$X(z) = \sum_{n=-\infty}^{\infty} x[n]\, z^{-n} \qquad (6.76)$$

More generally, $z = r\,e^{j\omega}$ is a complex number which can be represented in a complex plane. The relationship (6.76) is a particular case, where $|z| = 1$, that is, *the z-transform evaluated on a unit circle in the complex plane corresponds to the Fourier transform.*

Since $|X(z)| < \infty$ if:

$$\sum_{n=-\infty}^{\infty} |x[n]| \, |z|^{-n} < \infty \qquad (6.77)$$

the *region of convergence* (ROC) of the sum in eq.(6.76) consists of all values of z such that the inequality in (6.77) holds. Then if a given value of z is in the ROC, i.e. $z = z_1$, then all values on the circle defined in the complex plane by $|z| = |z_1|$ will also be in the ROC. This implies that, generally, the ROC assume the form of a ring comprised between the circles of radius $|z_1|$ and $|z_2|$, with $|z_1| < |z| < |z_2|$.

6.2.5 LTI Systems for Discrete-Time Signals

Linear Time-Invariant systems can be defined also for discrete-time signals (see [71]). Relationships analog to (6.40) and (6.42), valid for continuous-time signals, hold for sequences:

$$H(e^{j\omega n}) = \frac{Y(e^{j\omega n})}{X(e^{j\omega n})} \qquad (6.78)$$

$$y[n] = x[n] * h[n] \qquad (6.79)$$

A very important subclass of LTI systems is characterized by an input-output relationship which satisfies an Nth-order linear constant-coefficient difference equation of the form:

$$\sum_{k=0}^{N} a_k \, y[n-k] = \sum_{k=0}^{M} b_k \, x[n-k] \qquad (6.80)$$

Applying the z-transform to eq.(6.80) leads to

$$\sum_{k=0}^{N} a_k z^{-k} Y(z) = \sum_{k=0}^{M} b_k z^{-k} X(z), \qquad (6.81)$$

which can be rearranged in a form which is analog to relationship (6.78):

$$H(z) = \frac{Y(z)}{X(z)} = \frac{\sum_{k=0}^{M} b_k z^{-k}}{\sum_{k=0}^{N} a_k z^{-k}}, \qquad (6.82)$$

or, equivalently:

$$H(z) = \frac{b_0}{a_0} \cdot \frac{\prod_{k=1}^{M}(1 - c_k z^{-1})}{\prod_{k=1}^{N}(1 - d_k z^{-1})} \qquad (6.83)$$

The form (6.83) directly shows the position of *zeros* and *poles*[16] of the system function in the complex plane.

6.2.6 Numeric filters

In the context of analog signals, a filter is an electric network, or circuit, which alters (the frequency content of) an input signal x(t) such that to deliver an output signal $y(t)$. In the context of discrete-time signals, a numeric filter is a set of operations to be performed on a input sequence $x[n]$ in order to obtain the output sequence $y[n]$ (see eq.(6.79)).

An *ideal* filter lets the input signal unaltered over a certain range of frequencies, the *transmission band*, or *pass-band*, and suppresses it in the remnant portion of the

---

[16] Zeros and poles are solutions, respectively, of numerator and denominator of the system function expressed in the polynomial form (6.82) or (6.83).

frequency axis, the *attenuation band*, or *stop-band*. This implies an *infinite* slope of the filter attenuation profile, which is neither physically nor numerically realizable. Therefore, real filters present a *transition band*, which is the interval of the frequency axis given by the difference of the frequencies marking respectively the pass-band and the stop-band. In the transition band, the attenuation gradually changes from the value of the transmission band (ideally null) to the value of the attenuation band (ideally infinite). Diminishing the extension of transition zone means sharpening the amplitude profile of the filter's frequency response, which happens at the expense of realization costs (for assembled analog filters), or computational costs (for numeric filters), since the *filter order* (see after) increases. Therefore, depending on the current application, a compromise must be found, between transition rapidity and costs.

In the case of the OL SW, it was noticed that increasing the filter order beyond a certain limit (~ 15) implies numerical stability problems, since some *filter coefficients* (see after) get too close to zero (instability starts with coefficients values $O(10^{-17})$). This was not an issue of concern: filters of order comprised in the range 6 to 10 present a decay rate of the amplitude response in the range of $\sim$ 5- 10 $dB/decade$, which has proven to be well suitable for the OL data processing (see Ch. 7).

The numeric filters which will be considered belong to the class of the LTI systems, therefore can be described by a linear constant-coefficient difference equation of the kind of eq. (6.80). In order to represent the involved linear constant-coefficient difference equation by a block diagram, a more convenient form is used:

$$y[n] - \sum_{k=1}^{N} a_k\, y[n-k] = \sum_{k=0}^{M} b_k\, x[n-k] \qquad (6.84)$$

This equation can be rewritten as a recurrence formula for $y[n]$ in terms of a linear combination of past values of the output sequence and current and past values of the input sequence:

$$y[n] = \sum_{k=1}^{N} a_k\, y[n-k] + \sum_{k=0}^{M} b_k\, x[n-k] \ ,\qquad (6.85)$$

where the *order* of the filter is the greater of N or M, and the values a_k and b_k are the filter *coefficients*.

The value of Eq. (6.85) can be represented by the general block diagram shown in fig. 6.11 ([71]).

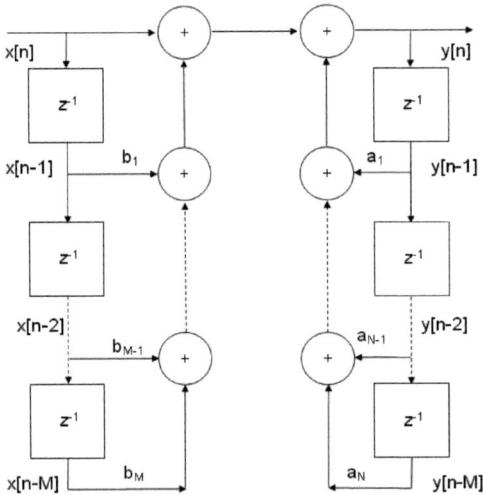

Figure 6.11: *Block diagram representation for a general N^{th}-order difference equation.*

Numeric filters are classified into two categories: the Finite Impulse Response (FIR)- and the Infinite Impulse Response (IIR) filters. As their names suggest, the basic difference lies in the length of the impulse response sequence. The answer to the question, when to use the one or the other type, depends on the current application.

Advantages and drawbacks of both filter categories must be considered, in order to find out which trade-off can best meet the project requirements.

Although only the IIR filter category has found application in the OL SW (see par. 7.1.5), a short overview of the characteristics of both kinds will be given, in order to provide argumentation for the choice effectuated.

FIR Filters and tapering windows

The simplest method of FIR filters design is called the *window method* ([71]). After selection of the desired frequency response $H(e^{j\omega})$, the corresponding (infinite) impulse response $h[n]$ is multiplied by a finite sequence of opportune form (window), the most common of which is the *rectangular* window $w[n]$, i.e. a unitary sequence of a certain length N. Supposing that the ideal filter $H(e^{j\omega})$ has a rectangular shape, its impulse response $h[n]$ will be of the type *sinc* (see par. 6.1.3), so that the new impulse response will be the product of the *sinc* function with a rectangular window. This corresponds in the frequency domain to the convolution of the Fourier transforms of the two functions, respectively a *rect* and a *sinc*. As a consequence, the obtained frequency response will be a "smeared" version of the ideal rectangular shape. The effect of windowing on the original filter is that of bringing oscillations in the original rectangular shape of the filter because of the side-lobes of the *sinc* function. Fig. 6.12 shows the analyzed example.

By writing the ideal frequency response as the Fourier transform of $h[n]$:

$$H(e^{j\omega}) = \sum_{n=-\infty}^{\infty} h[n] e^{-j\omega n}, \qquad (6.86)$$

defining the truncated impulse response as $h' = h[n] \cdot w[n]$, the new impulse response becomes:

$$H'(e^{j\omega}) = \sum_{n=-\infty}^{\infty} h'[n] e^{-j\omega n} = \sum_{n=-N/2}^{N/2} h'[n] e^{-j\omega n}, \qquad (6.87)$$

that is a *truncated* Fourier expansion of the original function $H(e^{j\omega})$. Therefore, the oscillations can be thought of as consequence of the *Gibbs' phenomenon* (see

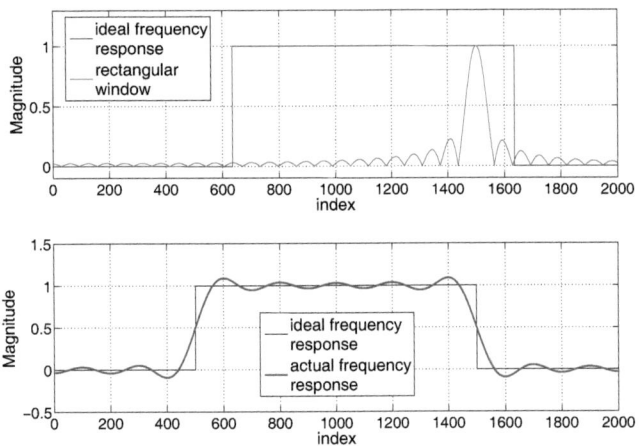

Figure 6.12: *Realization of a FIR filter by means of the windowing method with a rectangular window. Upper: ideal frequency response (blue curve), and Fourier transform of the window (red curve). Lower: red curve: result of the convolution of the ideal frequency response (black curve) with the Fourier transform of the window (red curve in the upper panel).*

par. 6.1.2). Analogously to what mentioned about the Fourier expansion, where increasing the number of considered harmonic terms does not reduce the error, but only confine the oscillations in a smaller interval around the discontinuity, extending the length of the window leads to more rapid oscillation with the same amplitude.

Adopting a different shape for the window, i.e. a window with softened transitions at the windows edges, reduce the hight of the side-lobes of the Fourier transform of the window, so that the effect of the oscillations can be possibly limited. The draw-back of this procedure is an enlargement of the main lobe in the frequency domain, thus a wider transition region at the points of discontinuity in the ideal frequency response. Therefore, the choice of the proper windows critically depends on the current application.

Figs. 6.13 and 6.14 show different types of tapering windows and their relative Fourier transforms. By comparing the frequency response of the various windows it is possible to see how an attenuation of the side lobes leads to a wider main lobe.

The method of the tapering windows is not only used in the context of the design of FIR filters: it finds application in each case in which it is necessary to smooth a sharp transition of a curve (analog signals) or a sequence (discrete-time signals) in one domain (i.e. time- or frequency domain) in order to control the side-lobes on the other domain. A typical application is the reduction of the *leakage*. As mentioned, the phenomenon originates from a not perfect matching of the fundamental period (the observation period) with the period of the analyzed harmonic (see 6.2.3), which, therefore, is not periodic in the time record. This causes discontinuities at the edges of the time record, when the acquired wave form is repeated in aperiodic pattern. The implementation of tapering windows would lead to a smoother edge transition ([2]).

In the frame of the spectral analysis performed in the OL SW, the DFT is executed on sets of samples of given length, which corresponds to an *implicit* windowing by means of the rectangular window. Since the looked-for information is the frequency of the spectral peak, sharpness of the main lobe was preferred to the attenuation of the side-lobes, so that the rectangular window resulted the most appropriate and no explicit tapering was adopted (see par. 7.1.4). (See also the observation at the end of par. 6.3.2).

FIR filters have the advantage of linear phase response and the draw-back of oscillating amplitude response, as it can be seen in fig. 6.15, which shows an example of FIR filter.

IIR Filters

An infinite impulse response, that is, an infinite sequence $h[n]$, is not practicable *per se*. Nevertheless, it can implemented through recursion, that is, bringing the output back to ingress (see eq.(6.85)).

The design of IIR filter starts from the selection of an *analog* filter which best approximates the numeric filter needed for the current application. Afterwards, the impulse response of the analog filter is sampled into a sequence $h[n]$ and transformed to $H(z)$, i.e., the frequency response is specified by means of an analog filter and then transferred to the z-domain. This method is called the *impulse invariance*. It is not advantageous for frequency ranges near the sampling frequency because

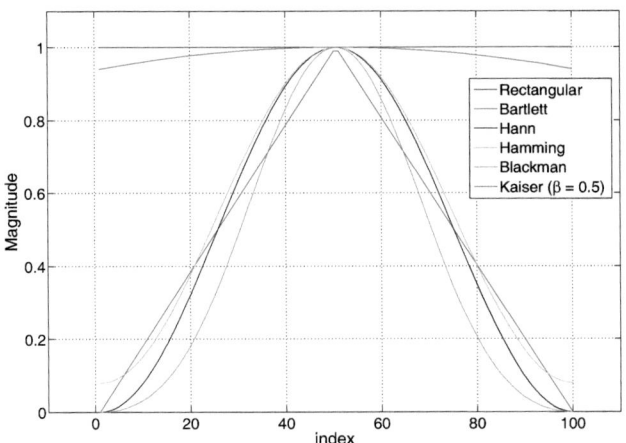

Figure 6.13: *Different types of tapering windows.*

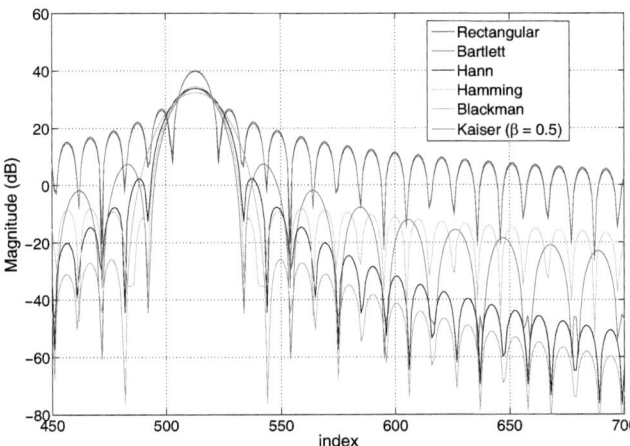

Figure 6.14: *Fourier transforms of the tapering windows shown in fig. 6.13.*

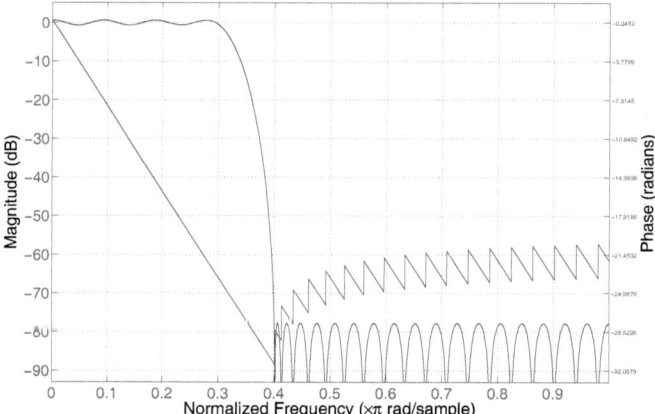

Figure 6.15: *Example of numeric low-pass FIR filter. The x axis represents the frequency in radians normalized to π. The left y axis shows the magnitude in dB (blue curve); the right y axis shows the phase in radians (green curve). Pass-band and stop-band were specified to 0.3 rad and 0.4 rad, respectively.*

of aliasing. This problem can be overcome with another design method, called the *bilinear transform*, in which a transformation, or pre-distorsion of the frequency axis of the analog filter is operated prior to covert the filter in the z-domain. This methods present other kinds of draw-backs, as, for instance, depending on applications, the computation complexity.

Opposite to FIR filters, IIR filters have the advantage of a linear ("almost" flat) amplitude response (attenuation in pass-band $\leq 0.1\,dB$) and the draw-back of non-linear phase response, as it can be seen in fig. 6.16, which shows an example of IIR filter.

The reason why this kind of filters was preferred for implementation in the OL SW is that it is possible to compensate the phase response of the filter directly in the application (see par. 7.1.5), whereas a linear attenuation can be more easily equalized than a "ripple" response, as the one of FIR filters. (In any case, the attenuation in pass-band is retained negligible).

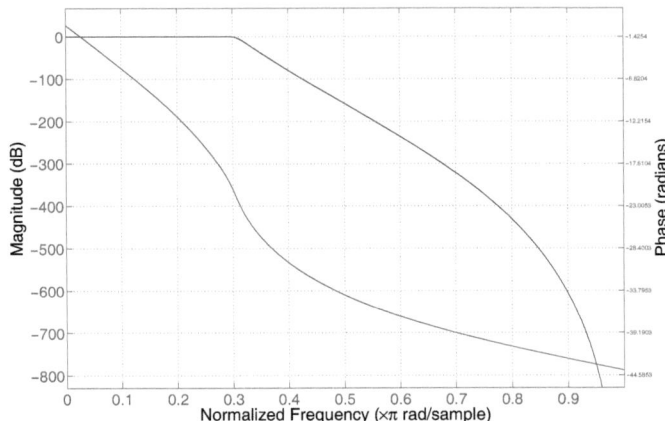

Figure 6.16: *Example of numeric low-pass IIR filter. The x axis represents the frequency in radians normalized to π. The left y axis shows the magnitude in dB (blue curve); the right y axis shows the phase in radians (green curve). Pass-band and stop-band were specified to 0.3 rad and 0.4 rad, respectively.*

6.3 Elements of Spectral Estimation[17]

In the following, a few generic aspects of the problem of spectral estimation will be considered, which are of significance for the developed *Open Loop* software package. In particular, the classical methods of spectral estimation, based on Fourier analysis, will be shortly introduced. As it will be dealt with discrete-time signals, the concepts presented for continuous stochastic processes will be extended to sequences.

"The general problem of spectral estimation is that of determining the spectral content of a random process based on a finite set of observations from that process". ([58]).

Given a complex wide-sense-stationary (WSS) random process $x[n]$, the power

---

[17] An extensive treatment of this subject can be found in [58]. See also [71].

spectral density (PSD) $P_{xx}(f)$ of such a process is defined as:

$$P_{xx}(f) = \sum_{k=-\infty}^{\infty} r_{xx}[k] \, e^{-j2\pi f k} \qquad -\frac{1}{2} \leq f \leq \frac{1}{2} \qquad (6.88)$$

where the frequency axis has been normalized to the sampling frequency, and $r_{xx}[k]$ is the autocorrelation function (ACF) of $x[n]$, defined as:

$$r_{xx}[k] = E(x^*[n]\, x[n+k]), \qquad (6.89)$$

whit "E" the expectation operator (see App. F.3). The relationship (6.88) is the Wiener-Kintchine definition of the PSD.

Since in eq.(6.88) the PDS depends on infinite samples of the ACF, the determination of the PDS is generally not possible. The spectral estimation problem can be therefore summarized as: "Based on the N contiguous observations $\{x[0], x[1], ..., x[N-1]\}$ of a single realization of a WSS random process, it is desired to estimate the PSD for $-\frac{1}{2} \leq f \leq \frac{1}{2}$". ([58]).

If, on the one side, it is self-evident that it is not possible to analyze an infinite set of samples, on the other side, the determination of N must be carefully pondered. Very often, the analyzed processes are not stationary (in the WSS sense), so that the ACF (6.89) does not only depend on k, but also on n, so that a PSD cannot be defined. In many cases the non-stationarity is not severe, so that the process can be defined *locally* WSS, that is, the variations of the ACF with n are "small" over the considered observation interval.[18] In this case a compromise must be met. On the one hand, a time-varying PDS would lead to smoothing or biasing in the estimation, but, on the other hand, a short data set would exhibit a large variability, due to a lack of averaging. The *bias-variance* trade-off is a basic characteristics of all spectral estimators. ([58]).

Another definition of the PDS, based on the Parseval theorem (see eq. (6.30))

---

[18]This is the case for the analyzed VeRa occultation data.

is:

$$P_{xx}(f) = \lim_{M \to \infty} \mathcal{E}\left\{ \frac{1}{2M+1} \left| \sum_{n=-M}^{M} x[n] e^{-j2\pi fn} \right|^2 \right\} \quad (6.90)$$

Under the assumption that the autocorrelation function r_{xx} decays sufficiently rapidly, the definition (6.90) is equivalent to the Wiener-Kintchine definition (6.88)

6.3.1 Estimation Theory

Some general definitions are introduced hereafter.

Estimation theory is concerned with the problem of estimating the values of parameters based on measured data that has a random component. An estimator $\hat{\boldsymbol{\theta}}$ attempts to approximate the unknown parameters $\boldsymbol{\theta}$ using the measurements $\mathbf{x}$ (bold letters indicate vectors).

There are some properties that a good estimator should posses. In the following, a scalar parameter θ is considered. If the estimator converges to the true parameter value(s) as the number of data points N increases, the estimator is said to be *consistent*. That is: $\lim_{N \to \infty} Pr\left(|\hat{\theta} - \theta| > \varepsilon\right) = 0$, where Pr denotes probability, and ε is a small positive number.

An indicator of the quality of an estimator is the *Mean Square Error* (MSE): ([58])

$$MSE = \mathcal{E}\{(\hat{\theta} - \theta)^2\} = \mathcal{E}\{(\hat{\theta} - \mathcal{E}\{\hat{\theta}\})^2\} + (\mathcal{E}\{\hat{\theta}\} - \theta)^2 \quad (6.91)$$

The first term is the *variance* of the estimator, whereas the second term is the squared of the *bias* of the estimator. For an unbiased estimator should be:

$$\mathcal{E}\{\hat{\theta}\} - \theta = 0 \quad (6.92)$$

6.3.2 Classical Spectral Estimation

Classical spectral estimation was used in the frame of the statistical error analysis in order to evaluate the statistics of noise (see par. 8.1.1).

In the following, ergodicity (see App. F.3.1) is assumed for the considered processes, so that a single realization of the process is sufficient to determine the ACF.

The method of the periodogram

The *periodogram spectral estimator* is based on the definition of PDS given in eq.(6.90). Since it is not possible to dispose either on an infinite number of samples or on an *ensemble* of realizations, the method of the periodogram omits the limit and the expectation operators in eq.(6.90), so that the PSD estimate is given by:

$$\hat{P}_{PER}(f) = \frac{1}{N} \left| \sum_{n=0}^{N-1} x[n] \, e^{-j2\pi fn} \right|^2 \qquad (6.93)$$

The estimate of the PSD could be performed also by starting from the Wiener-Kintchine definition of eq. (6.88)). In this case, given a realization of the process $x[n]$, due to the finite number N of samples at disposal, applying eq. (6.88) to the data would not yield the true PDS, $P_{xx}(f)$, but:

$$\hat{P}(f) = \sum_{k=-(N-1)}^{N-1} r_{xx}[k] \, e^{-j2\pi fk} \qquad (6.94)$$

For the same reason, the autocorrelation function can not be exact, as it has to be calculated from a finite number of samples, i.e., the N samples of the current realization of the process $x[n]$. That is, only a biased estimate $\hat{r}_{xx}[k]$ of the ACF is available, where:

$$\hat{r}_{xx}[k] = \begin{cases} \dfrac{1}{N} \displaystyle\sum_{k=-(N-1)}^{N-1} x^*[n]\, x[n-k] & \text{for } k = 0, 1, ..., N-1 \\ \hat{r}_{xx}^*[-k] & \text{for } k = -(N-1), -(N-2), ..., -1 \end{cases} \qquad (6.95)$$

It can be demonstrate that the periodogram estimate of the finite length sequence $x[n]$, $(n = 0, ..., N-1)$, given in eq.(6.93) is equivalent to the Fourier transform of the estimated autocorrelation sequence $\hat{r}_{xx}[k]$, that is:

$$\hat{P}_{PER}(f) = \sum_{k=-(N-1)}^{N-1} \hat{r}_{xx}[k]\, e^{-j2\pi fk} \qquad (6.96)$$

The expected value of the periodogram is: ([58])

$$\mathcal{E}\{\hat{P}_{PER}(f)\} = \sum_{k=-(N-1)}^{N-1} \mathcal{E}\{\hat{r}_{xx}[k]\}\, e^{-j2\pi fk} =$$

$$= \sum_{k=-(N-1)}^{N-1} \left[\dfrac{N - |k|}{N}\right] r_{xx}[k]\, e^{-j2\pi fk} = \mathcal{F}\{w_B[k] r_{xx}[k]\} =$$

$$= \int_{-1/2}^{1/2} W_B(f - \xi)\, P_{xx}(\xi)\, d\xi, \qquad (6.97)$$

where $W_B(f)$ is the Fourier transform of the Bartlett- or triangular window:

$$w[k] = \begin{cases} 1 - \dfrac{|k|}{N} & \text{for } |k| \le N - 1 \\ 0 & \text{for } |k| > N - 1 \end{cases} \qquad (6.98)$$

that is:

$$W_B(f) = \dfrac{1}{N}\left[\dfrac{sin(\pi f N)}{sin(\pi f)}\right]^2 \qquad (6.99)$$

The relationship (6.97) implies that the *average* periodogram is the convolution of the true PDS with the Fourier transform of a Bartlett window, yielding on the average a smoothed version of the true PSD. Thus the periodogram is generally *biased* for finite data record but unbiased as $N \to \infty$, since:

$$\lim_{N \to \infty} \mathscr{E}\{\hat{P}_{PER}(f)\} = P_{xx}(f) \tag{6.100}$$

For the case of white noise, the estimator is unbiased even for finite N, as the product of the Bartlett window by a Dirac delta function (that is, the ACF of the noise), would result in the ACF itself. Furthermore, it can be seen that the periodogram of white noise has χ^2 statistics. In fact, considering that each point of the periodogram is given by $|X[k]|^2 = X[k] \cdot X^*[k]$, $k = 0, ...N-1$, where $X[k]$ is the kth element of the DFT of the real noise sequence $x[n]$, by remembering eq. (6.73) it follows that:

$$\begin{aligned}
X[k] \cdot X^*[k] &= \sum_{n=0}^{N-1} x[n] \, e^{-j\frac{2\pi}{N}kn} \cdot \left(\sum_{n=0}^{N-1} x[n] \, e^{-j\frac{2\pi}{N}kn} \right)^* = \\
&= \sum_{n=0}^{N-1} x[n] \, e^{-j\frac{2\pi}{N}kn} \cdot \sum_{n=0}^{N-1} x[n]^* \, e^{j\frac{2\pi}{N}kn} = \\
&= \sum_{n=0}^{N-1} x[n]^2
\end{aligned} \tag{6.101}$$

From the relationship (6.101) follows that each sample of the power spectrum equals the sum of the square of N *normal distributed, independent* random variables, which, therefore, is χ^2 distributed ([74]).

Differently from the expected value, increasing the number of samples N does not reduce the variance of the periodogram, which, therefore, proves to be an *not consistent* estimator. This is due to the lack of an average operation. Since the periodogram is based on a single realization of the process, the values of the points of the periodogram will show a certain amount of variability for each given realization.

It can be shown that:

$$var[\hat{P}_{PER}(f)] \simeq P_{xx}^2(f), \qquad (6.102)$$

from which results that the variance of the estimator is a constant independent of N. This makes the periodogram an *unreliable* estimator, since its standard deviation is as large as its mean, which is approximately the quantity to be estimated. If several independent realizations of the examined process are available, the variance of the estimator can be reduced by *averaging* the single periodograms. Given K independent realizations of the process, the *averaged periodogram estimator* is defined as:

$$\hat{P}_{AVPER}(f) = \frac{1}{K} \sum_{m=0}^{K-1} \hat{P}_{PER}^{(m)}(f) \qquad (6.103)$$

In practice, independent data set are seldom, so that a common approach is to segment the N samples in K non-overlapping blocks of length L and average the resulting periodograms. This reduces the variance of the estimate at the expenses of the bias, since the shorter length of the single blocks results in a shorter Bartlett window (see eq.(6.98)), thus in a wider main lobe of its transform. In this case, a technique which is useful is the *prewhitening* of the data, since, as already mentioned, in the case of white Gaussian noise the periodogram is unbiased for any length of the analyzed data segment.

In some cases, it could be advantageous to apply a data window to the sequence to process in order to reduce sidelobes which could mask eventual weak signals which are spectrally close to stronger ones. Anyway, it can be shown that, *if the data consist of one complex sinusoidal signal*[19] *embedded in white Gaussian noise, then the optimal frequency estimator is the periodogram with no data windowing*[20].

---

[19]or several complex sinusoids, spaced more then f_s/N apart in frequency (where N is the number of points, and f_s the sampling frequency), that is, the frequency resolution bin.

[20]This is the case for the analyzed VeRa occultation data in the regions where no multipath propagation is involved.

The Blackman-Tukey Estimator

This estimator is based on the definition (6.96) of the periodogram. The adopted principle is to minimize the contribution of those samples in the estimated ACF that are more imprecise, i.e. the samples at higher lag (displacement), because of the fewer averaged lag products. This can be done by means of a weighting window, $w[k]$, which is a real sequence. The estimates would be:

$$\hat{P}_{BT}(f) = \sum_{k=-(N-1)}^{N-1} w[k]\,\hat{r}_{xx}[k]\,e^{-j2\pi fk} \qquad (6.104)$$

The sequence $w[k]$, which performs the weighting of the ACF estimator, is called *lag window*. It has the properties:

- $0 \leq w[k] \leq w[0] = 1$;
- $w[-k] = w[k]$;
- $w[k] = 0 \quad \text{for } |k| > M$;

where $M \leq N - 1$. Due to the listed properties, the relationship (6.104) can be written as:

$$\hat{P}_{BT}(f) = \sum_{k=-M}^{M} w[k]\,\hat{r}_{xx}[k]\,e^{-j2\pi fk} \qquad (6.105)$$

The relationship (6.105) is called the *Blackman-Tukey spectral estimator* (BT estimator). It is equivalent to the periodogram if $w[k] = 1$ for $|k| \leq M = N - 1$. It is also called a *weighted covariance* estimator: the weighting of the ACF reduces the variance of the estimator at the expense of increasing the bias (wider mainlobe of the Fourier transform of the lag window).

From the relationships (6.96), and (6.105), observing that

$$\hat{P}_{BT}(f) = \mathscr{F}\{w[k]\,\hat{r}_{xx}[k]\} = \int_{-\frac{1}{2}}^{\frac{1}{2}} W(f - \xi)\,\hat{P}_{PER}(\xi)\,d\xi, \qquad (6.106)$$

for the average of the BT estimator is:

$$\mathscr{E}\{\hat{P}_{BT}(f)\} = \int_{-\frac{1}{2}}^{\frac{1}{2}} W(f-\xi)\, \mathscr{E}\{\hat{P}_{PER}(\xi)\}\, d\xi\,, \qquad (6.107)$$

which, when evaluated for large data records, leads to:

$$\mathscr{E}\{\hat{P}_{BT}(f)\} \simeq \int_{-\frac{1}{2}}^{\frac{1}{2}} W(f-\xi)\, P_{xx}(\xi)\, d\xi\,, \qquad (6.108)$$

since the periodogram is unbiased when $N \to \infty$. As for the periodogram, the mean is a smeared version of the true PDS. It can be shown that:

$$var[\hat{P}_{BT}(f)] \simeq \frac{P_{xx}^2(f)}{N} \sum_{k=-M}^{M} w^2[k]\,, \qquad (6.109)$$

As an example, if the Bartlett window is chosen (in order to compare with the periodogram case):

$$var[\hat{P}_{BT}(f)] \simeq \frac{2M}{3N} P_{xx}^2(f) \qquad (6.110)$$

It can be seen that choosing an appropriate window, both the type and the length, is a central issue in classical spectral estimation.

Chapter 7

"Open Loop" Data Processing Software

7.1 Basic principle

In this chapter are described the digital processing techniques which are implemented in the software package for the evaluation of the Radio Science occultation data in the "Open Loop" format (OL). Although optimized for the specific conditions imposed by the Venus atmosphere, the processing strategy outlined hereafter is based on a generalized approach which allows application to any kind of data recorded in OL mode, for example also for cometary data. In the following the case of the VEX Radio Science occultation experiment at Venus is considered.

In order to define a convenient approach to the problem the expected characteristics of the signal to be received are considered. The carrier signal transmitted by the spacecraft ideally consists of a single tone of constant amplitude A_0 at the nominal carrier frequency f_c. Considering the interaction of the microwave with interposed media and also the thermal noise contribution of the receiver, the signal at the receiver output can be written in the most general form as[1]:

$$s(t) = [A_0 + \Delta A(t)] \cdot cos\{2\pi[f_c + \Delta f_{Doppler}(t) + \Delta f_V(t) + \\ + \Delta f_E(t) + \Delta f_P(t) + f_{PN}(t)] \cdot t\} + n(t) \qquad (7.1)$$

---
[1]Here the attenuation of *free-space loss* and the amplification of the receiver chain are not considered.

where:

- $\Delta A(t)$ expresses amplitude variations (supposed to be due only to the Venus atmosphere);

- $\Delta f_{Doppler}(t)$ is the time-varying frequency shift imposed by the classical Doppler effect (straight-line Doppler)[2];

- $\Delta f_V(t)$ represents the frequency variation due to the Venus ionosphere and atmosphere;

- $\Delta f_E(t)$ represents the frequency variation due to the Earth ionosphere and atmosphere (troposphere)[3];

- $\Delta f_P(t)$ is the frequency variation imposed by interplanetary plasma (solar wind);

- $\Delta f_{PN}(t)$ describes the effects of phase noise and aging of the frequency source (see App. D.1);

- $n(t)$ is the thermal noise of the receiver (assumed to be white Gaussian noise).

As mentioned in Ch.5, the parameter of interest are the amplitude- and frequency variation which are induced by the Venus atmosphere and ionosphere, that is, $\Delta A(t)$ and $\Delta f_V(t)$. Those quantities are contained in the amplitude and frequency of the received signal. Whereas it is possible to estimate the motion-induced Doppler frequency shift by means of ephemeris calculation and reconstructed orbit data, to calibrate the effects of the Earth environment almost completely by means of in-situ measurements and look-up tables ([3]), and to compensate the effect of interplanetary plasma by means of differential Doppler (see par. 2.3.2), the effects of the frequency instability of the spacecraft signal and of the thermal noise at the

---

[2]This is the contribution to the total frequency shift of the received signal which is due to the relative motion between S/C and G/S, considering free space propagation (vacuum). The effects of the crossed medium are not considered in this quantity.

[3]The Earth troposphere causes the received signal to experience an additional path delay which ranges from $\sim$ 2-2.5 m for large elevation angle of the tracking antenna (zenith), to $\sim$ 20-28 m at a 5° elevation angle, depending on air pressure, temperature, and humidity ([45], [40]). Due to the relative motion of the satellite and G/S, the the total phase path length of the radio signal shows time dependency which leads to a measurable Doppler frequency shift of $\sim$ 40 mHz at X-Band and $\sim$ 10 mHz at S-Band. The additional frequency shift caused by the Earth ionosphere is about one order of magnitude smaller than the aforementioned quantities ([45], [40]). See also par. 8.2.1.

receiver are inseparable from the received signal and dictate the sensitivity of the experiment.[4]

It is convenient to consider the received signal after removal of the carrier frequency and motion-related Doppler frequency shift and calibration of the effects of the Earth atmosphere- and ionosphere. Equation (7.1) becomes:[5]

$$s(t) = A(t)\,cos[2\pi f(t) \cdot t] + n(t) \qquad (7.2)$$

where the variable frequency $f(t)$ reflects the frequency variations induced by the transected media (i.e., the ionosphere and atmosphere of the planetary body) as well as the instability of the frequency source and $A(t)$ represents the signal amplitude, whose variations depend on the crossed media, too. A proper choice of the length of the observation window "T" allows considering the signal $s(t)$ as "quasi-stationary", that is, slowly varying over time intervals having the same order of magnitude as the reciprocal of the frequency at a particular time t_0 in the selected time interval T ([64]). For this assumption to hold it must be:

$$T(t)|_{t_0} \simeq f^{-1}(t)|_{t_0} \qquad (7.3)$$

This is a central issue for the outlined processing approach, as explained in the following.

The strategy for the evaluation of the OL occultation data relies on iterative processing which allows a progressive reduction of the signal bandwidth. At each iteration step, an estimation of the signal frequency takes place. The time series provided by the estimation is used to compute the phase of a complex exponential function of unit amplitude which, in turns, is used for numerical down-conversion of the signal itself: the signal samples are multiplied by the samples of the generated

---

[4]In this regard the frequency instability of the H_2-Maser used for the down-conversion on ground is negligible as its Allan deviation is two orders of magnitude smaller than the Allan deviation of the USO. The same ratio applies for thermal noise, whose standard deviation is comparable with the USO instability.

[5]The effects of interplanetary plasma possibly lying on the radio path between S/C and G/S can be at first neglected.

complex function in order to reduce signal dynamics and to reposition the signal spectral content in the center of the frequency span. Ideally this process should result in a sinusoidal signal of zero frequency already after the first down-conversion.[6] The uncertainty of the estimate causes the signal to retain residual dynamics, that is, a time-varying value of its frequency which is defined as *frequency residual*.

The various signal frequency estimates, performed at each iteration step in order to obtain the relevant down-conversion signal, are based on Fourier analysis. A Fast-Fourier-Transform algorithm (FFT, see par. 6.2.3) is applied to subsequent sets of signal samples covering a time interval T, over which amplitude and frequency of the signal are considered being approximately constant, as mentioned in relation to eq. (7.3). Since the signal is not perfectly stationary, it is affected by a slow residual drift in the considered time interval T, which, along with superimposed thermal noise, is responsible for the inaccuracy of the frequency estimate.

The chosen iterative approach has the advantage of reducing the signal dynamics step by step, which allows progressive extension of the observation time T. This, in turn, implies increased resolution power of the FFT, and, therefore, larger values of SNR in each resolution cell (frequency bin). Since signal degradation increases with penetration depth (due to defocusing- and absorption effects, see par. 5.2.3), the SNR improvement produced by the recursive approach allows the recovery of information from increasingly low altitudes.

Typically, the selected values of T vary from 1 s at the beginning of the processing to 10 s at the end of the processing, which is usually completed within three-four iterations, depending on occultation depth.

Each FFT-based estimate of the frequency is smoothed by a proper algorithm (see par. 7.1.2), in order to obtain a signal which is suitable for the down-conversion process of the following stage.

At the end of each processing step, the down-converted samples undergo numeric low-pass filtering, which allows sample decimation, a considerable advantage in terms of computation velocity and memory allocation.[7]

---

[6]Not considering amplitude variations, which result in some narrow-band spectral content placed around zero frequency.

[7]In principle, the implementation of low-pass filtering at the end of each iteration step has the only advantage of sample decimation. It has no influence on the frequency estimate of the next step, since this is based on the FFT, where the noise bandwidth is dictated by the length of the observation window and it holds $1/T^{(n)} < f_c^{(n)}$, because $1/T^{(n)}$ shall not exceed the residual signal dynamics sd_n (see eq. (7.4)). (T is the duration of the observation window in the FFT, f_c is the

In order not to compromise the information contained in the current frequency residual, three different contributions to the actual bandwidth occupation have to be taken into account when assessing the cut-off frequency of the filter. They are:

1. the absolute maximum frequency shift Δf_n reached by the frequency residual w.r.t. to zero;

2. the residual signal dynamics sd_n, given by the change rate of the frequency residual;

3. the uncertainty on the performed estimate, ϵ_n.

It must be:

$$f_{cut}^{(n)} \geqslant BW_n = \Delta f_n + sd_n + \epsilon_n \qquad (7.4)$$

where $f_{cut}^{(n)}$ is the cut-off frequency of the low-pass filter and BW_n is the bandwidth occupation of the signal, both considered at the nth processing step.

The frequency variations over time, which are observed in the frequency residuals obtained in the various processing steps, let us consider the down-converted, received signal as being PM/FM modulated (see App. A). Therefore, the Carson formula is used for the evaluation of the signal bandwidth BW_n. In this way, the residual signal dynamics sd_n is considered to be the frequency of a fictive signal which modulates the carrier and, therefore, expressed in $[Hz]$.

In the praxis, a margin of $\sim 20\%$ of the estimated signal bandwidth is considered in the selection of the cut-off frequency of the filter.

Typically, the frequency residuals considered in the first processing iteration step (after removal of straight-line Doppler frequency shift) present a maximum

---

cut-off frequency of the filter, and apex n represents the n-th iteration step). Furthermore, since the frequency profile is built up by looking at the maximum in each frequency spectrum (see after, par.7.1.4), the filter has also the effect to optimize the estimate in the middle of the pass, when the signal strength is comparable with the noise power. In this case the signal peak can be easily masked by peaks of noise. The effect of the filter is to reduce the number of the frequency bins which "compete" for the maximum, thus increasing the probability of the right frequency bin to be chosen (point "D" in fig. 7.6.)

frequency shift comprised in a range of 60-80 Hz (deep occultation passes), with maximum signal dynamics of $\sim 1\,Hz/s$. Since eq. (7.3) holds, for large values of the SNR ($SNR > 10\,dB$) the estimation error ϵ_n is contained within the width of the frequency resolution bin of the FFT spectrum. This leads to $\epsilon_1 = 1\,Hz$, since the observation window T at the first iteration stage is normally set to be 1 second. Therefore, for a typical deep-occultation pass, a cut-off frequency $f_{cut}^{(1)} = 100\,Hz$ is selected for the low-pass filter of the first processing stage.

The thermal noise at the receiver and the frequency stability of the USO set the maximum accuracy which is attainable by the progressive reduction of the noise bandwidth and signal dynamics, thus providing a limit for the number of iterations. The process should stop when the values of the estimated frequency are comparable with the inaccuracy of the estimate itself, that is, error bars of the estimate are comprised between 50% and 100%. Typically, the analysis is accomplished within three to four iteration steps, as already mentioned. The frequency profiles obtained in each iteration step are stored in order to use them in the reconstruction of the total frequency shift at the end of the processing.

Once the iterations are completed, the signal obtained after the last down-conversion is evaluated in the time domain in order to obtain *instantaneous* values of the frequency. This requirement is not attainable by means of the FFT, as the calculated spectra result from averaging over the whole observation period T. At the end of the processing all frequency estimates from the performed iteration steps are added to the instantaneous frequency in order to obtain the resulting media-induced frequency shift.

As long as the relationship (7.4) holds, the accuracy of the frequency estimation depends on the accuracy of the time domain estimate and not on the various estimates obtained by means of the FFT during the iterative procedure. This is due to the fact that all performed down-conversions serve exclusively to reposition the signal to the center of the spectrum and to reduce signal dynamics. They have to be regarded merely as "arbitrary" shifts which are "undone" at the end of the processing by adding them to the time domain estimate. The power of the received signal is calculated in time domain as well.

The outlined processing strategy has the advantage of preserving an arbitrary time resolution (in principle the full sample rate!) independently of the integration time. By means of the described progressive reduction of the signal dynamics, it is possible to cope with low values of the SNR without reducing the time resolution by averaging, since the final frequency evaluation is performed in the time domain as instantaneous frequency. This is not the case for CL data. As a matter of fact, in the CL mode the phase information of the incoming signal is extracted from the error signal of the PLL, which implies a fixed value of receiver bandwidth (see also 5.3). This implies that the only way to cope with a degrading SNR is to increase the integration time of the phase values, which happens at expense of the spatial resolution in the probed atmosphere.

The followed approach can be regarded as an enhancement of the so-called "Radio-Holographic" techniques (see par. 7.2.2), in which the frequency profiles are directly obtained from the power spectrum of the signal after removal of the straight-line Doppler and the expected atmospheric frequency shifts. The improvement brought by the OL software is provided by the iterations and the time analysis, as these increase SNR and time resolution at the same time.

A special handling is needed for the data blocks affected by *multipath* propagation. The topic is discussed in the paragraph 7.2.

7.1.1 Implementation of predicted expected frequency shift in the OL software

Before the iterative processing starts, some preliminary operations are performed on the delivered samples. As described in Ch.5, upon reception at the ground station, the signal is down-converted to IF ($70\,MHz$), digitized, and numerically mixed-down by a coarse frequency predict, which approximately removes the straight-line (vacuum) Doppler frequency shift. As a consequence, the provided signal samples are still affected by some residual motion-related signal dynamics, and, of course, by the effects of the interposed media (i.e. the Venus atmosphere- and ionosphere). The first operation performed on the samples is to correct the down-conversion performed at the G/S by means of the accurate straight-line predict provided by the RSS (see Ch.3). Since the information about the applied down-mix function is saved and delivered in auxiliary data, it is possible to calculate the difference between the G/S predict and the RSS predict and numerically mix the samples by the computed error function. The purpose of the removal of the residual straight-line Doppler shift is to have a signal whose dynamics show exclusively the effect of the investigated media.[8]

As described in 5.3, the raw data delivered by the ground station for processing consist in two set of orthogonal samples, the *in-phase* sequence, represented by $I[n]$ and the *quadrature* sequence, represented by $Q[n]$, where brackets represent index notation, as usual for discrete-time signals.[9] As mentioned in par. 5.3, both sequences can be combined in the complex notation:

$$s^+[n] = I[n] + j\,Q[n] = A[n]\,e^{j\psi[n]} \qquad (7.5)$$

where:

$$\begin{cases} A[n] = \sqrt{I^2[n] + Q^2[n]} \\ \psi[n] = \arctan\left(\frac{Q[n]}{I[n]}\right) \end{cases} \qquad (7.6)$$

and $s^+[n]$ is the *analytic signal* associated to $s[n]$, with $s[n] = A[n]\,cos(\psi[n])$ (see par. 6.1.7).

After phase correction this sequence becomes:

$$s[n] = A[n]\,e^{j\phi[n]} \qquad (7.7)$$

---

[8]The effects of the Earth atmosphere- and ionosphere are compensated at a later stage of the data processing and not in the OL software. In the following, however, it is assumed that they have been already removed.

[9]In the following, the terms "signal" and "sequence" will be used equivalently to denote discrete-time signals, if not differently specified. Furthermore, signal spectra will be handled as spectra of sampled analog signals, rather than as spectra of numeric sequences, difference being the dimension of the abscissa, which is still $[Hz]$ in the former case, whereas it becomes dimensionless in the domain of the sequences and is given in radians (see 6.2). Therefore, the first spectral replica of the sampled signal (the base-band spectrum), comprised in the interval $[-f_s/2, f_s/2[$, where f_s is the sampling frequency, will be considered as the power density spectrum of the original analog signal.

where

$$\phi[n] = \psi[n] - (\phi_{str}[n] - \phi_{GS}[n]) \tag{7.8}$$

where ϕ_{str} is the phase shift induced by the straight-line Doppler effect[10] (see fig. 7.1). It was calculated by interpolating the frequency values of the predict, $f_{pr}[i]$, on the time instant given by the time stamps of the samples and by multiplying the obtained new series $f[n]$ by the sampling period Δt and adding up all values (numerical integration):

$$\phi_{str}[n] = \sum_{l=1}^{n} f_{pr}[l] \cdot \Delta t \tag{7.9}$$

Similarly, ϕ_{GS} was obtained from the values of the frequency of the local oscillator at the ground station, $f_{LO}[i]$, which are contained in auxiliary files, as mentioned in 5.3:

$$\phi_{GS}[n] = \sum_{l=1}^{n} f_{LO}[l] \cdot \Delta t \tag{7.10}$$

The first estimation of the received signal frequency after correction of classical Doppler effects is not based on the data but on a ray tracing calculation ([79]) (see fig. 7.2) using a reference atmosphere derived from the *Magellan Venus* mission ([62]). This allows reaching a degree of accuracy on the first frequency estimate which would not be attainable from the data itself because the noise bandwidth has not yet been reduced at the beginning of the precessing. This in turns implies

---

[10]The subtlety of including the correction term in brackets in eq. 7.8 renders the implementation of the calculus: in principle the predict of the G/S must be "undone" and the correct predict must be applied on the restored "sky-frequency" (frequency of the received signal, as it would be measured at the antenna terminals, before down-conversion). In practice is not possible to bring the samples to oscillate at the sky-frequency $(O(10^9)\,[Hz])$, as the sampling rate $(10^5\,[s^{-1}])$ is too low. If one were to add the phase of G/S predict to the phase of the samples, one would obtain a phase ambiguity of multiples of 2π, that is, one would see the base-band portion of the aliased sky-frequency. Instead, what is conceptually correct is to calculate the phase difference of the two predicts (straight-line predict from the RSS and G/S-predict, as in (7.8)), since the result is a *slowly* varying phase which can be used to correct the phase of the samples.

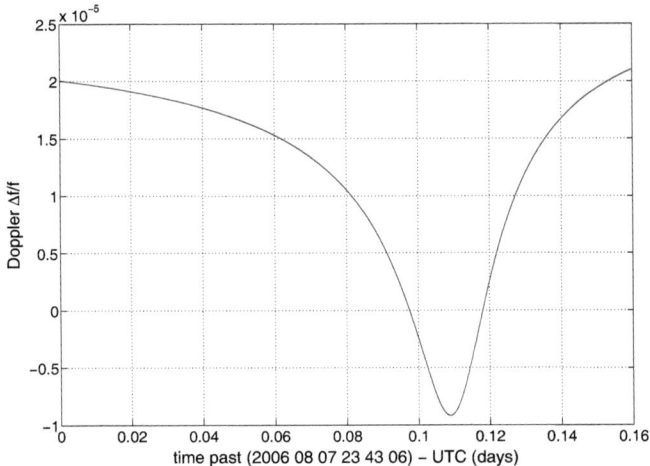

Figure 7.1: *Prediction of the expected frequency shift induced by the relative motion of transmitter and receiver in vacuum (classical Doppler effect, or straight-line Doppler) for the VEX-VeRa occultation pass on DoY220-2006, orbit #109. The values are relative, being expressed as $\Delta f/f$ (where f is the transmitted frequency), thus valid for both transmission frequencies (X- and S-band). The absolute frequency shift is calculated in the OL SW by multiplying the predict by the USO-driven transmitter frequency. This one was obtained by interpolation over a look-up table, filled up with values from observations carried out throughout the mission during occultation seasons (see par. 4.5.1). For the X-Band D/L reference value of $8.4\,GHz$, an absolute maximum straight-line shift of $\sim 70\,kHz$ is obtained at $t \sim 0.109$ days.*

the possibility to dramatically reduce the signal dynamics in a pre-processing stage, which has the advantage to optimize memory allocation and computational speed. Typically, the implementation of the atmospheric predict implies a decimation factor of the magnitude order of 10^2. This allows to retain at most one sample per record (records contain 87 samples, see 5.3), which greatly simplify the SW structure.

A pre-processing step is performed ahead of the iterative procedure in order to evaluate the bandwidth of the signal $s_0[n]$ after removal of straight-line Doppler frequency and atmospheric predict, where

$$s_0[n] = s[n]\, e^{-j\phi_{atm}[n])} = A[n]\, e^{j(\phi[n] - \phi_{atm}[n])} \qquad (7.11)$$

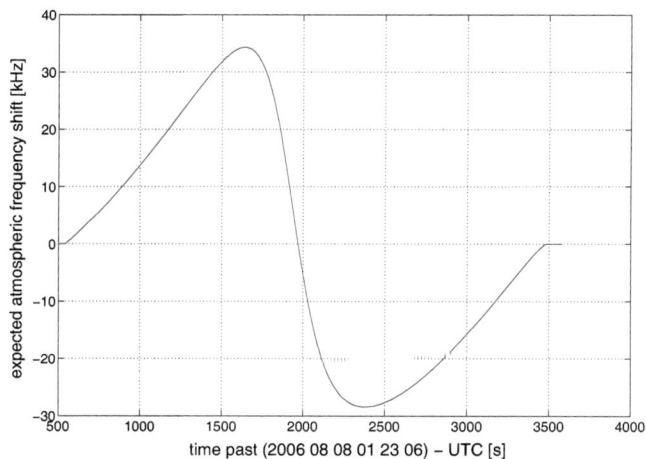

Figure 7.2: *Prediction of the expected frequency shift induced by the Venus atmosphere on the VEX X-Band RF carrier signal for the VEX-VeRa occultation pass on DoY220-2006, orbit #109. Time dependent variations of the bending angle of the rays in the Venus atmosphere cause Doppler frequency shifts on the received signal. The maximum frequency excursus varies during the occultation season, as it depends on the geometry of the occultation: it is small for limb occultations and increases for larger scanned portions of the atmosphere. The maximum is in coincidence of the diametric pass, or diametric occultation, when the projection of the S/C track on ground reaches its maximum length. The selected pass was couple of orbits past the diametric occultation (see par. 5.1).*

with $s[n]$ defined in eq. (7.7), and ϕ_{atm} being the phase calculated from the atmospheric predict by the same procedure as in eqs. (7.9) and (7.10).

Figures 7.3 and 7.4 show block diagrams representing the pre-processing stage and the first iteration step of the OL software. (The down-conversion process is repeated in the step #1, as it not possible to store the down-converted samples in one memory structure. It was retained more convenient running the same piece of SW twice rather than writing new code and managing the write/read process of memory arrays). In the mentioned- and following block diagrams which describe the OL SW structure, green cylinders represent read/write storage units, PC symbols correspond to the processing of input quantities (incoming arrows). Arrows departing from PC

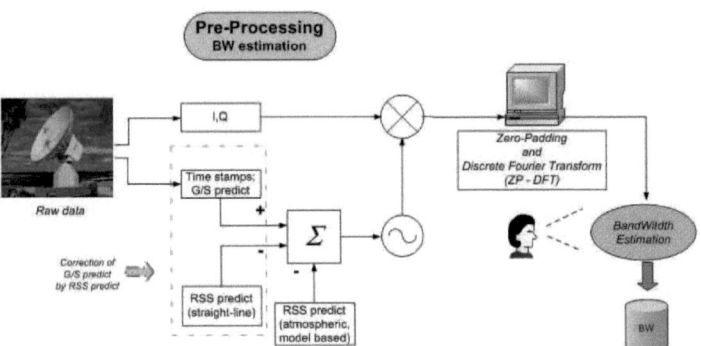

Figure 7.3: *Block diagram showing the pre-processing stage of the OL software. The RSS straight-line predict is employed in order to correct the inaccuracy of the G/S predict. The prediction of the atmospheric-induced frequency shift is use to considerably reduce signal dynamics at the beginning of the processing (decimation factor $> 10^2$), in order to optimize memory allocation and calculation speed. The pre-processing is performed in order to estimate the bandwidth of the first frequency residual, so that the cut-off frequency of the low-pass filter of the first stage can be determined.*

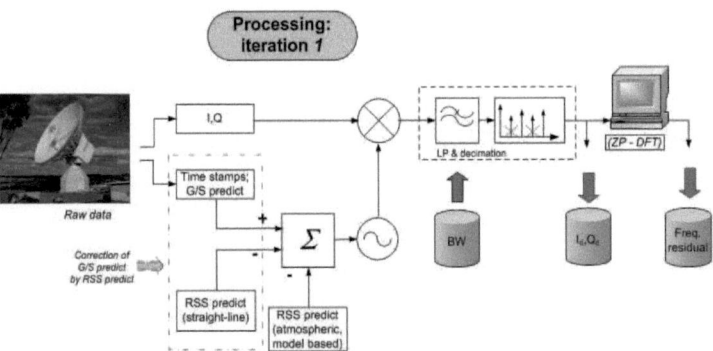

Figure 7.4: *Block diagram showing the first processing step of the iterative processing procedure. After removal of straight-line Doppler frequency and expected atmospheric frequency shift, the samples are low-pass-filtered and decimated. The cut-off frequency of the filter is determined on the basis of the bandwidth evaluation occurred in the pre-processing stage.*

symbols represent results of computation. Oscillators and mixers are characterized by the conventional symbols and represent numeric operations, i.e., the computation of a sequence of frequency values and the multiplication of complex exponential sequences, respectively.

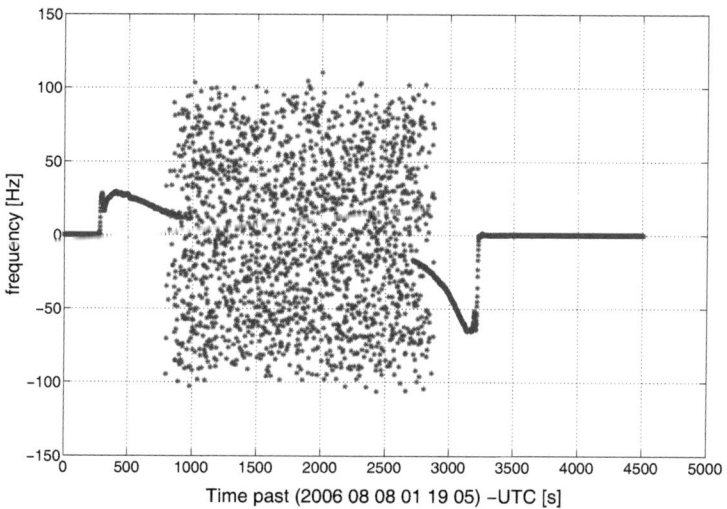

Figure 7.5: *VEX-VeRa occultation pass on DoY220-2006, orbit #109. Frequency vs. time profile from FFT-based analysis after the first processing iteration step. Expected straight-line Doppler frequency shift and atmosphere-induced frequency shift (shown in fig. 7.1 and fig. 7.2, respectively) have been removed. The down-converted samples were low-pass filtered and decimated accordingly to the Nyquist criterion. The filter cut-off frequency was 80 Hz. The presence of noise beyond the interval ±80 Hz is due to the finite slope of the applied filter (amplitude response decay: 5 dB/decade, see par 6.2.6). Each frequency point derives from FFT performed on sample sets covering a time span of 1 s; the time indication of each point is referred to the center of the interval. Time resolution: 1 s; frequency resolution cell (width of the FFT frequency bin): $\sim 40\,mHz$.*

Figure 7.5 shows the frequency residual evaluated on the basis of the Fourier analysis after the removal of the expected Doppler frequency shift, both straight-line and atmosphere-induced, showed in fig. 7.1 and fig. 7.2, respectively.

In the following, the various operations carried out in the generic n^{th} iteration step will be considered in some details, along with the procedure which computes the total frequency shift from the partial results of the single iteration steps.

7.1.2 Single iteration steps: signal frequency evaluation, polynomial fit and storage of coefficients

Figure 7.6 shows the block diagram of the generic n^{th} iteration step. The $(n-1)^{\text{th}}$ frequency residual from the previous iteration step is smoothed in order to obtain a noiseless function suitable for the mix-down process of the current iteration step. The smoothed curve is then fitted by a 3rd-order polynomial and the computed polynomial coefficients are stored for the build-up of the total frequency shift at the end of the iteration process (point "A"). The voltage samples from the previous step are mixed-down by means of the calculated polynomial (point "B"). After evaluation of the bandwidth occupation, the down-converted samples are low-pass filtered, decimated, and stored for the next iteration step (point "C"). From the samples a new frequency profile is computed, which will serve as mix-signal in next stage (where it will be smoothed and fitted) (point "D").

The frequency residuals have a typical signature which exhibits different dynamics along the pass, so that the smoothing action must be locally adjusted. This was achieved by calculating a set of curves characterized by different depths of the smoothing action, each one covering the whole duration of the pass. A segment of each smoothed curve was then selected and the single segments were joined together in order to cover the whole pass. The difference among the implemented degrees of smoothing must be kept within a certain limit, in order to avoid strong discontinuities in the resulting curve. Small discontinuities are flattened by a mild smoothing on the obtained curve. In the final curve the averaging effect gradually increases from the beginning to the middle of the pass and decreases from the middle to the end. Figure 7.7 shows the frequency residual after the first iteration step and the superimposed smoothed profile. Figure 7.8 shows the frequency residual after the second iteration step: the smoothed profile of the previous step (red curve in fig. 7.7) was used as down-conversion signal. The central portion of the smoothed profile (noisy data) serves as connection between the initial- and final parts of the residual curve, rather than as reliable estimation of the frequency, as the SNR is too low. Nevertheless, as the signal is expected to lower its dynamics in the middle of the pass (see par. 7.2.1), signal could show up in further processing stages, since the

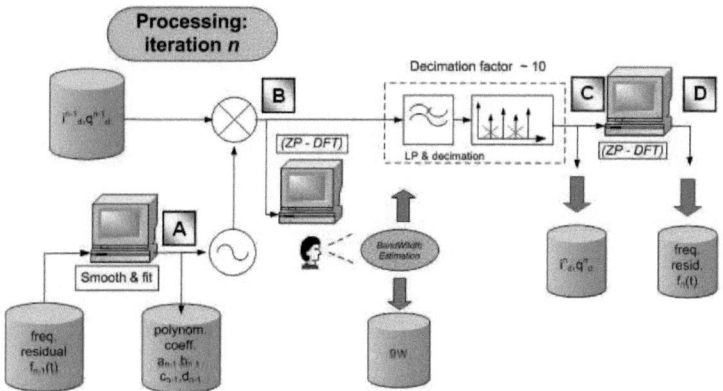

Figure 7.6: *Block diagram showing the nth processing step of the iterative processing procedure. The frequency residual from the previous iteration step is smoothed and fitted by a 3rd-order polynomial and the computed polynomial coefficients are stored for the build-up of the total frequency shift at the end of the iteration process (point "A"). The voltage samples from the previous step are down-converted by means of the calculated polynomial (point "B"), and, after evaluation of the bandwidth occupation, low-pass filtered, decimated, and stored for the next iteration step (point "C'). From the samples a new frequency profile is computed, which will serve as mix-signal in next stage (where it will be smoothed and fitted) (point "D").*

noise bandwidth is progressively reduced. In this regard, the smoothing action on the frequency estimate is intensified in the middle of the pass, in order to prevent the mix-signal to oscillate after noise and to possibly drive the weak signal toward the wrong direction.

7.1.3 Computation of last frequency residual and received signal power from time samples and construction of the total frequency shift

After completion of all iteration steps, estimates of signal intensity and frequency are obtained as time series from the analysis of the signal base-band components in

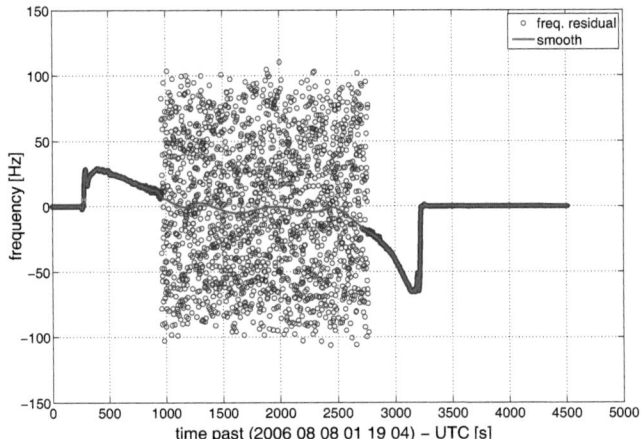

Figure 7.7: *VEX-VeRa occultation pass on DoY220-2006, orbit #109. Smoothed frequency residual after the first processing iteration step. Blue points: original data; red curve: smoothed data. The frequency residual was obtained by means of Fourier analysis on the down-converted, low-pass filtered samples (see fig. 7.5); the removed straight-line- and atmospheric predicts are shown in fig. 7.1 and fig. 7.2, respectively. (Interpretation and comments are given in text).*

the time domain. In digital form:

$$\begin{cases} P[n] = \varrho^2[n] + \epsilon_p \\ f[n] = \dfrac{\Delta\varphi[n]}{\Delta n} + \epsilon_f \end{cases} \quad (7.12)$$

where:

$$\begin{cases} \varrho[n] = \sqrt{a^2[n] + b^2[n]} \\ \varphi[n] = \arctan\left(\dfrac{b[n]}{a[n]}\right) \end{cases} \quad (7.13)$$

with $a[n]$ and $b[n]$ the recorded samples of the signal base-band components; ϵ_P and ϵ_f represent the uncertainties on the signal intensity and signal frequency, respectively. For the retrieval of instantaneous frequency and power, the residual

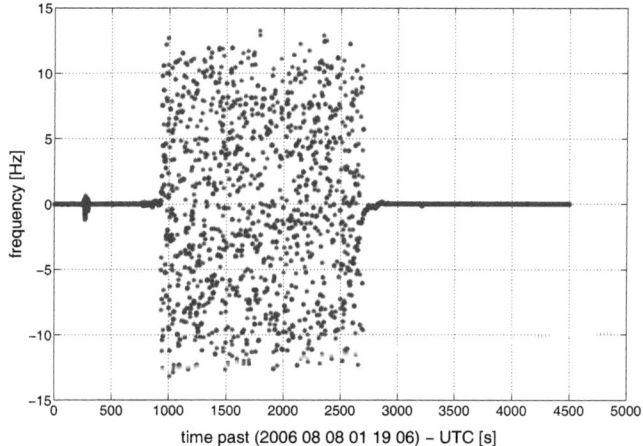

Figure 7.8: *VEX-VeRa occultation pass on DoY220-2006, orbit #109. Frequency vs. time profile from FFT-based analysis after the second processing iteration step. The samples were down-converted by frequency estimation performed on the output data of the previous step (see fig. 7.7), low-pass filtered (cut-off frequency: $10\,Hz$) and decimated accordingly to the Nyquist criterion. Each frequency point derives from FFT performed on sample sets covering a time span of $2\,s$; the time indication of each point is referred to the center of the interval. Time resolution: $0.1\,s$; frequency resolution cell (width of the FFT frequency bin): $\sim 2.5\,mHz$.*

signal dynamics make a time domain analysis more accurate than a FFT-based approach. This observation is based on the intrinsic feature of a Fourier spectral analysis of being an integral measurement over a certain time span. Hereby, past and future information are averaged in each spectrum (typically, $T = 10\,s$ at the last iteration step) whereas a time based analysis provides phase information at the selected sample rate ($\Delta t = 0.1\,s$ or better.) The maximum time resolution selected should be compatible with the extension of the Fresnel zone in the atmosphere ($\sim$ 60- to $70\,m$ in vertical direction for deep sounding altitudes ($h \sim 40\,km$), see par. 5.2.4).

Figure 7.9 shows the last processing step, which takes place after completion of the iterative procedure. The frequency functions applied in the down-conversion process of each iteration step are calculated from the stored polynomial coefficients on the time basis of the residual samples and added to the instantaneous frequency profile in order to obtain the total atmospheric-induced frequency shift. The received signal power profile is computed from the residual samples in the time domain.

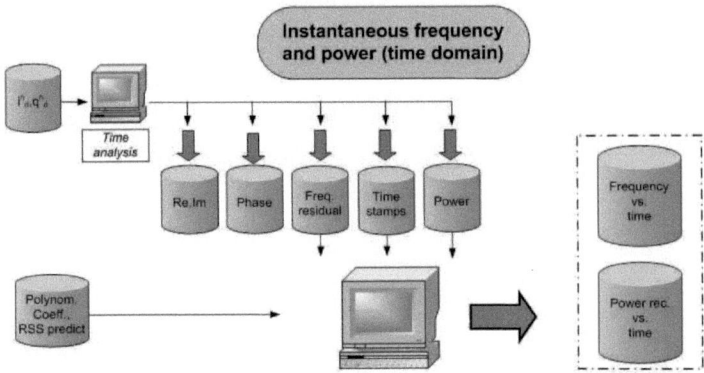

Figure 7.9: *Block diagram showing the last processing step of the OL SW. The reduced samples are evaluated in the time domain and instantaneous frequency- and power time series are made available. By means of the stored polynomial coefficients it is possible to calculate the subtracted frequency residuals at the times given by the time stamps of the remaining samples in order to build up the total frequency shift.*

The effect of the progressive reduction of the signal dynamics and the correspondent low-pass filtering is clearly visible in figs. 7.10 to 7.13. This example, taken from the VEX-VeRa occultation pass on 8th-Aug-2006 (DoY 220), shows the time profile of the received signal power calculated from the time samples at the end of each processing step. The bandwidths of the applied numeric filters correspond to $80\,Hz$, $10\,Hz$, $5\,Hz$, $0.5\,Hz$, respectively from iteration step 1 to iteration step 4 (the last one). Each processing step introduce an improvement in the SNR, so that the curve of the received power, which contains the effects of absorption and defocusing, emerges from background noise. The effects of *multipath propagation* (see par. 7.2) are visible during the ingress- and egress phase, where superimposition of more signals let the power of the received signal exceed the nominal value in vacuum.

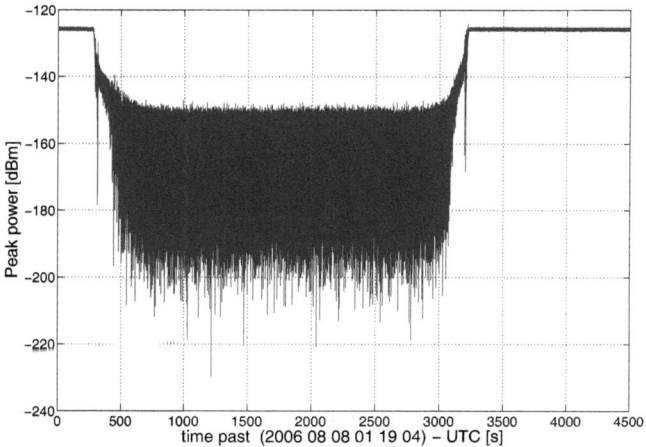

Figure 7.10: *VEX-VeRa occultation pass on DoY220-2006, orbit #109. Peak power of received signal from time series. Iteration step 1. Low-Pass filter cut-off frequency: 80 Hz.*

7.1.4 FFT-based frequency estimation

Due to the implicit rectangular windowing applied when executing the FFT over a finite number of samples, the obtained power density spectra (periodograms) will be of the form $sinc^2$, where $sinc(x) = \dfrac{sinx}{x}$, centered around the current value of the signal frequency ([64]).

The frequency estimation at each processing step, that is, a frequency profile vs. time over the whole data set, is obtained by looking at the frequency of the peak in each spectrum. In order to improve the time resolution, the FFT is applied in "sliding" mode, that is, the signal samples of each observation window partly overlap with the samples of the next window. Typically, a time resolution of $0.1\,s$ is implemented. This implies an overlapping of the observation windows of 90%.

As the prime information needed is the location of the spectral peak on the frequency axis, no explicit windowing is executed on the time samples (see par. 6.2.6). Time samples were zero-padded prior to perform FFT in order to increase the frequency accuracy when locating the maximum within the main lobe (see par. 6.2.3).

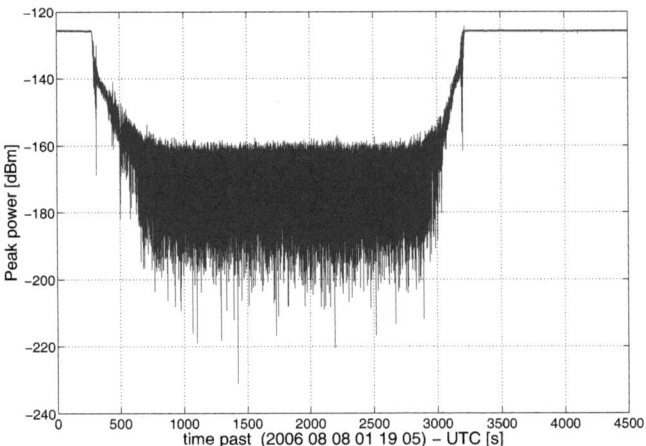

Figure 7.11: *VEX-VeRa occultation pass on DoY220-2006, orbit #109. Peak power of received signal from time series. Iteration step 2. Low-Pass filter cut-off frequency:* $10\,Hz$.

The criterion used to determine the integration time T is based on the requirement that each spectrum maintain a $sinc^2$ shape. This limits the maximum allowed frequency shift of the carrier in the observation time T to a certain percent of the main lobe width. If this limit is exceeded, the $sinc^2$ shape is warped into the spectrum of a chirp signal. This consideration sets a relationship between observation time T and frequency shift rate of the signal (signal dynamics). Setting the maximum allowed frequency shift over T to 50% of the main lobe width f_{ML} yields:

$$\left.\frac{df(t)}{dt}\right|_{MAX} = \frac{f_{ML}}{2T} \qquad (7.14)$$

where $\left.\frac{df(t)}{dt}\right|_{MAX}$ reflects the maximum signal dynamics. Considering that the main lobe width f_{ML} corresponds to the reciprocal of the observation time interval

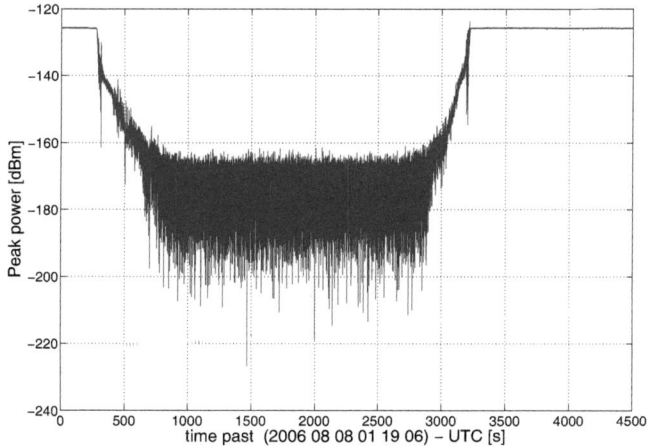

Figure 7.12: *VEX-VeRa occultation pass on DoY220-2006, orbit #109. Peak power of received signal from time series. Iteration step 3. Low-Pass filter cut-off frequency: 5 Hz.*

T yields:

$$\left.\frac{df(t)}{dt}\right|_{MAX} = \frac{1}{2T^2} \tag{7.15}$$

Figure 7.14 shows an example spectrum taken from the VEX-VeRa occultation pass occurred on DoY220-2006, orbit #109. The FFT was performed on a 1s sample set ($T = 1\,s$), taken at $t = 1\,s$ (center of interval) after beginning of recording, before the signal entered the atmosphere, after down-conversion (removal of the expected straight-line frequency shift and predicted atmospheric frequency shift), low-pass filtering and decimation. The spectrum shows the expected $sinc^2$ shape, thus confirming the hypothesis quasi-stationarity. This is not the case for a sample set of the same pass taken at the same processing stage at $t = 285\,s$ (center of interval) after beginning of recording, when the signal was well within the atmosphere (shown in 7.15). The discrepancy of the adopted atmospheric model from the real atmosphere causes the signal to drift in the observation window, causing a chirp-like spectrum.

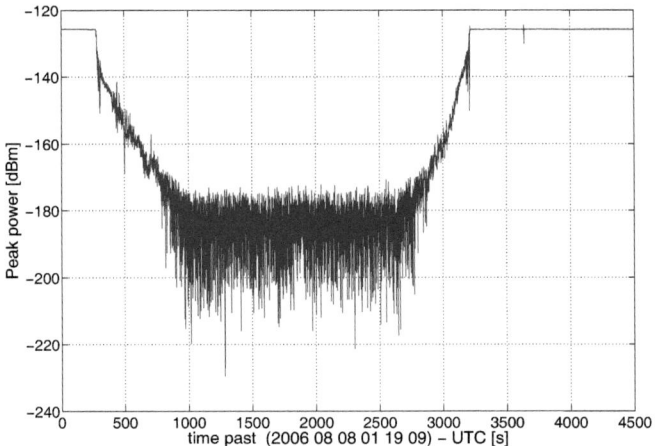

Figure 7.13: *VEX-VeRa occultation pass on DoY220-2006, orbit #109. Peak power of received signal from time series. Iteration step 4. Low-Pass filter cut-off frequency: 0.5 Hz.*

The observation window must be diminished in order to unambiguously track the carrier signal.

7.1.5 Design of numeric low-pass filters

As mentioned in Ch.6, the action of a filter on a transiting signal is that to alter the shape, that is, the spectral content of the signal itself, accordingly to the needed application. This operation is represented mathematically as the convolution of the signal with the pulse response of the filter (see par. 6.2.5), that is:

$$y(t) = x(t) * h(t) = \int_{-\infty}^{+\infty} x(\tau)h(t-\tau)d\tau \qquad (7.16)$$

where x(t) is the input signal, h(t) is the pulse response of the filter, and y(t) is the output signal. The character "*" represents the convolution operator.

Provided a "suitable" digital form of the filter pulse response, the same operation

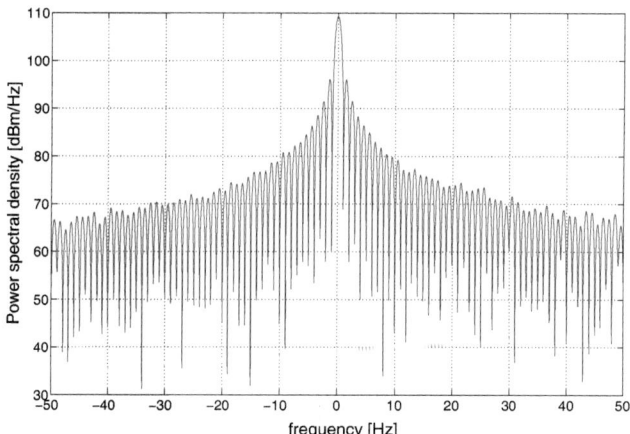

Figure 7.14: *VEX-VeRa occultation pass on DoY220-2006, orbit #109. Power density spectrum from FFT. Observation time $T = 1\,s$. Frequency resolution cell (width of the FFT frequency bin): $\sim 40\,mHz$. Data taken at $t = 1\,s$ after beginning of recording (center of interval), before the signal entered the atmosphere, after removal of the expected straight-line frequency shift and predicted atmospheric frequency shift (shown in fig. 7.1 and fig. 7.2, respectively).*

can be applied to digital signals. In this case the integral operator is substituted by summation, as the values are defined at discrete times:

$$y[n] = x[n] * h[n] = \sum_{-\infty}^{+\infty} x[k]\, h[n-k] \qquad (7.17)$$

As discussed in Ch.6, the discrete pulse response $h[n]$ can be computed by means of two different strategies, which lead to two different categories of numeric filters: the *infinite impulse response-*, or IIR filters, and the *finite impulse response-*, or FIR filters. Defining $X(z)$ and $H(z)$ as the z-transforms of the input sequence and of the filter pulse response, respectively, the output sequence in the domain of the z variable is expressed by:

$$Y(z) = X(z) \cdot H(z) \qquad (7.18)$$

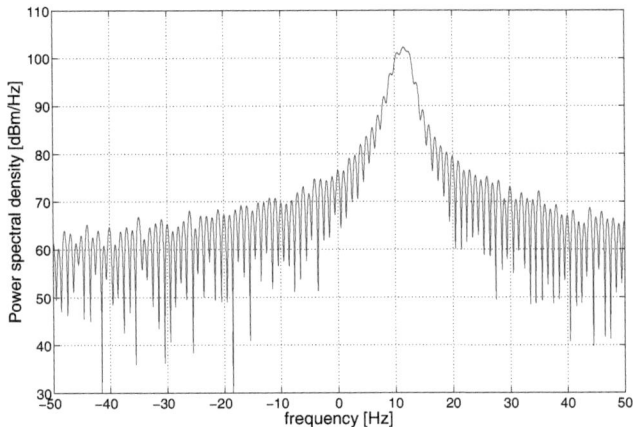

Figure 7.15: *VEX-VeRa occultation pass on DoY220-2006, orbit #109. Power density spectrum from FFT. Observation time $T = 1\,s$. Frequency resolution (width of the frequency bin): $\sim 40\,mHz$. Data taken at $t = 285\,s$ after beginning of recording (center of interval), after removal of the expected straight-line frequency shift and predicted atmospheric frequency shift (shown in fig. 7.1 and fig. 7.2, respectively), when the signal was well within the atmosphere. The discrepancy of the adopted atmospheric model from the real atmosphere causes the signal to drift in the observation window, causing a chirp-like spectrum. The observation window must be diminished in order to unambiguously track the carrier signal.*

where $H(z)$ is a rational expression whose coefficients depends on the desired filter characteristics (see par.6.2.6).

The computation of the coefficients of the numerical low-pass filter used at each processing stage was supported by MATLAB dedicated routines. The IIR filter category was preferred to the FIR kind because of its linear residual amplitude attenuation in pass band which is easier to account for when compared with the amplitude fluctuations which are characteristic of FIR filters (see figs. 6.15 and 6.16 in par.6.2.6). Still, IIR filters have the drawback of non-linear phase response which would affect the processed signal by irreversible distortion, if not compensated. The problem could be solved by cascading the low-pass filter block and reversing the output sequence after each block (anti-causal processing is always possible in no-real-time applications as the data are recorded and stored). Referring to (7.18), from the properties of the z-transform, the reversed output sequence in the z-domain will have the form:

$$Y(1/z) = X(1/z) \cdot H(1/z) \tag{7.19}$$

which, after further application of the low-pass filters yields:

$$\hat{Y}(z) = Y(1/z)H(z) = X(1/z)H(1/z)H(z) \tag{7.20}$$

Reversing the last sequence in the time domain yields for the z-transform:

$$\tilde{Y}(z) = X(z)H(z)H(1/z) \tag{7.21}$$

which, considered for $|z| = 1$, that is $z = e^{j\omega}$ yields:

$$\tilde{Y}(e^{j\omega}) = X(e^{j\omega})H(e^{j\omega})H(e^{-j\omega}) = X(e^{j\omega})H(e^{j\omega})H^*(e^{j\omega}) =$$
$$+ X(e^{j\omega})|H(e^{j\omega})|^2 \tag{7.22}$$

completely removing the phase response of the filter. Theoretically, the filtered data should be compensated for the amplitude variations introduced by the filter. Since these lie within a fraction of dB, they are at first retained negligible.

As mentioned in par. 6.2.6, filters of order comprised in the range 6 to 10 with decay rate of the amplitude response $\sim$ 5-10 $dB/decade$ where implemented in the OL SW.

7.2 Multipath propagation

During the analysis of the data, superimposition of more carrier signals has been regularly observed at certain points of the data set for a few tens of seconds (see figs. 7.18 and 7.19). These spectral features show up shortly after the beginning of the atmospheric penetration, in the "ingress" phase, and -symmetrically- shortly after that the signal re-emerges from the deep atmospheric layers on the other side of the planet, in the "egress" phase. These time segments correspond to a height of the scanned atmosphere comprised between 50-70 km, in correspondence with the region of the cloud layer ([93]).

This phenomenon is caused by small-scale structures (i.e. undulations, or sinusoidal features) of the refractive index, associated with temperature inversion layers in the cloud region (70 - 50 km). The departure of the vertical gradient of the refractive index from the monotonic profile of the background structure lets different "rays"[11] (typically, an odd number of them) contained in the transmitting antenna beam to converge in the focus located at the receiving antenna of the G/S.

Fig. 7.16 is intended to illustrate the principal of multipath propagation using the example of the Earth atmosphere. In this case, a GPS satellite (left-hand side of the figure) set at infinite distance, transmits a set of parallel rays, each one sent at a different epoch. A Low-Earth-Orbit (LEO) satellite (right-hand side of the figure) receives the transmitted signal along its orbit. The non-monotonic gradient of the refractivity causes some rays to cross the optical paths of others. Therefore, these rays are received in a chronological order which differs from the one of the transmission epochs, and interfere with the "current" ray being received. In the case of planetary occultation at Venus, the transmitter is located at the S/C and the receiver is represented by the G/S on the Earth.

The received signal is composed by the coherent summation of all signals whose optical path has the Earth in the focus for a given time instant. As a consequence, the amplitude and phase of the single components is not directly observable (see fig. 7.17). In order to compute the value of the refractive index of the resolution cell (Fresnel zone in the atmosphere) scanned by each of the convergent rays, one should be able to obtain the bending angle of each single path, which implies the recovery of the Doppler shift of each individual signal in the multipath zone.

Figures 7.18 and 7.19, show time series of the spectra obtained from VEX occultation data affected by multipath propagation. Straight-line Doppler shift, and expected atmospheric shift were removed by down-conversion. Since the atmospheric frequency predict is calculated for a unique, "expected" signal, the spectral components of the down-converted received signal will have different dynamics, with relative phases $\phi_m(t) - \phi_k(t), (k = 1...n)$, (subscript m denotes the signal assumed

---
[11]The phenomenon is here described from a *geometric optic* perspective.

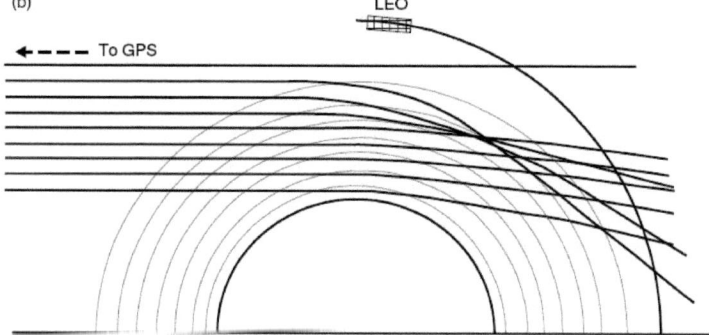

Figure 7.16: *Representation of multipath reception in an Earth occultation experiment. On the left-hand side, a GPS satellite (not shown), transmits rays at different epochs. On the right-hand side, a Low-Earth-Orbit (LEO) satellite receives the transmitted signal throughout its orbit. The non-monotonic gradient of the refractivity causes some rays to cross the optical paths of others. Therefore, these rays are received in a chronological order which differs from the one of the transmission epochs, and interfere with the "current" ray being received.* (Adapted from [69]).

as the nominal one), depending on the size of the refractivity perturbation and on the local defocussing ([69]). The spectral distance amongst the overlapped received signals from the multipath region ranges from a few tenth of Hertz to zero, making the recovery of the single components over the whole multipath event by means of a FFT-based analysis not possible, due to the intrinsic conflict between time- and frequency resolution in the Fourier transform. A different approach will be presented in paragraph 7.2.3.

7.2.1 The "Multi-track" routine

The processing strategy outlined in previous sections fails when dealing with data originating from multipath propagation, as it presupposes a single, "slowly-varying" carrier signal to be detected. Forcing the algorithm to track only the strongest spectral peak lead to discontinuous profiles (see fig. 7.20), since the power of each of the overlaid signals fluctuates with time, depending on the crossed portion of atmosphere, but also on the interference itself, when the signals become closer in the spectrum. Furthermore, the assumption that the strongest signal is the "right"

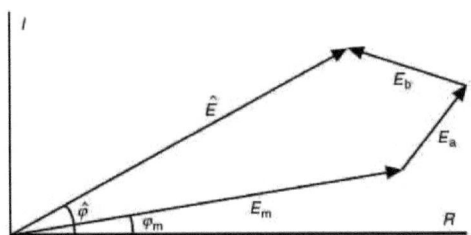

Figure 7.17: *Vectorial sum of received signal components for a "multipath" event. Each vector is represented as a phasor with amplitude* $\mathbf{E}$ *and phase* ϕ. *The resulting phasor is* $\hat{\mathbf{E}}^{j\hat{\phi}}$. *Subscript m denotes the "main" signal, that is, the ray which is expected to propagate in absence of refractivity anomalies.* (Adapted from [69]).

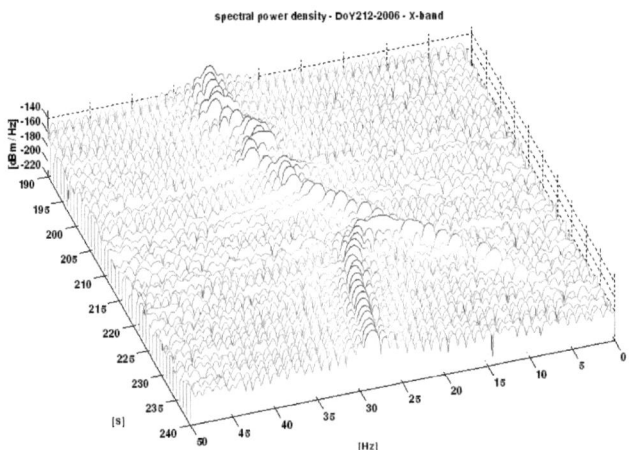

Figure 7.18: *VEX-VeRa occultation pass on DoY212-2006, orbit #101. Waterfall diagram showing the time sequence of power density spectra of received signal in the multipath region during the ingress phase. An odd number of carrier signals is expected (three, or more), but it is not possible to clearly distinguish more than two spectral peaks, mostly because of the inadequacy of the FFT to obtain proper time- and frequency resolution concurrently.*

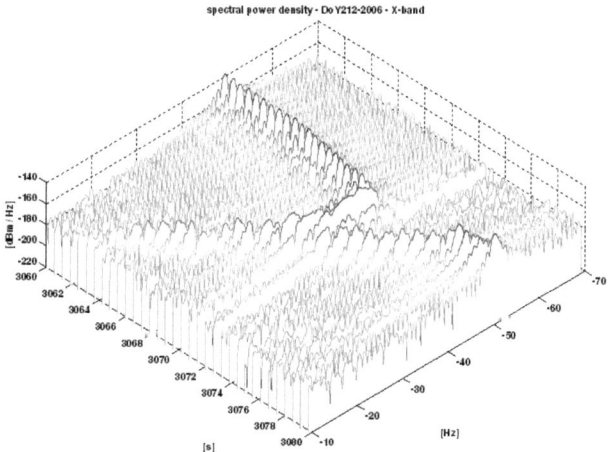

Figure 7.19: *VEX-VeRa occultation pass on DoY212-2006, orbit #101. Waterfall diagram showing the time sequence of power density spectra of received signal in the multipath region during the egress phase. An odd number of carrier signals is expected (three, or more), but it is not possible to clearly distinguish more than two spectral peaks, mostly because of the inadequacy of the FFT to obtain proper time- and frequency resolution concurrently.*

one does not match the physical phenomenon originating multiple reception, as already mentioned. Nevertheless, it was necessary to process the residuals in such a way to provide the down-conversion stage with a mixing-signal which is continuous and shifts a (hypothetical) main carrier signal toward the center of the spectrum, being aware that the frequency profile obtained at the end of the iterative process requires further elaboration in the multipath regions. To this purpose, it was noticed that considering a certain number of secondary maxima and inflection points would contribute to soften the transition(s): the underlying adopted model is that of a "nominal" carrier signal hidden amongst the interfering secondary signals. Therefore, a "multitrack" routine was developed in order to look for stationary points (maxima and inflection points) in each spectrum and record their frequencies in a matrix as functions of time (see fig. 7.21). Next, the routine scans the matrix along the time dimension in order to build up a frequency vs. time profile. The first selected frequency value is the first entry of a selected frequency time track (the one which corresponds to the main lobe of the stationary signal before the interference); in each of the following steps, the routine scans all frequency entries which correspond to the time $t_n + 1$ and selects the frequency value $f_n + 1$ which is next to the previous one (f_n), regardless to which track it belongs (secondary maximum or inflection point). Figure 7.22 shows how the developed routine reduces discontinuities in the frequency profile. Residual discontinuities have to been removed by the smoothing stage, in order to can feed the down-conversion stage with a continuous signal.

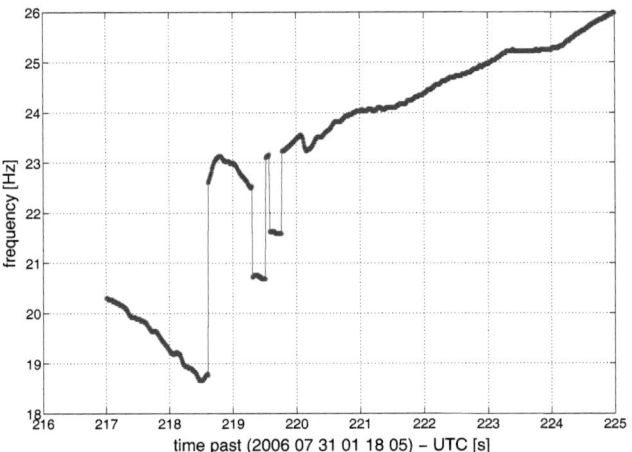

Figure 7.20: *VEX-VeRa occultation pass on DoY212-2006, orbit #101. Frequency vs. time profile, Data segment from the multipath region. The standard maximum-search routine provides a broken profile as it "locks" on different "rays" from spectrum to spectrum, depending on the relative strength of the received signals.*

Extension of atmospheric sounding towards low altitudes

The "multi-track" routine was also applied to the data segments characterized by a low value of the SNR ($SNR \leq 0\,dB$, considered in $1\,Hz$-bandwidth, which is the usual frequency resolution adopted in the first iteration processing step), that is, at the point where the received signal power (continuously decreasing after the

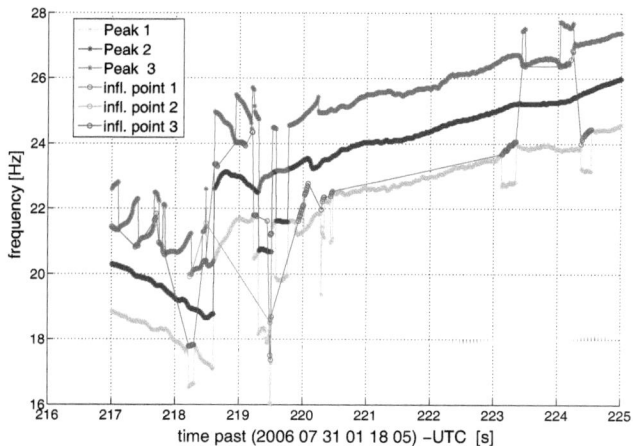

Figure 7.21: *VEX-VeRa occultation pass on DoY212-2006, orbit #101. Frequency vs. time profiles provided by the "multi-track" routine applied on a data segment from the multipath region.*

ingress of the spacecraft carrier signal in the Venus atmosphere) reaches the noise floor and -symmetrically- at the point where it comes out of the noise floor, when the transmitted carrier signal emerges from the other side of the planet[12]. Here the signal power is comparable with the average noise power contained in the frequency resolution cells of the FFT. By selecting the absolute maximum in each spectrum there is a given probability that the chosen frequency bin contains the signal (detection probability, Neyman-Pearson criterion). If one were to make some *a priori* assumptions on the expected signal, the detection probability could be increased.

By applying the "multi-track" routine to the noisy data, a certain number of FFT points is selected in each spectrum, besides the absolute maximum. These values will be called "secondary maxima". As already mentioned about the application of the "multi-track" routine (see par. 7.2.1), the algorithm builds a "database" of frequency vs. time profiles which, therefore, will be as many as the specified number of secondary maxima. As an example, there will be a frequency profile

---

[12]The level of attenuation which makes the incoming signal comparable with the noise floor is, of course, relative to the adopted spectral resolution (reciprocal of the duration T of the FFT observation window). Since, typically, for the first processing iteration step is $T = 1\,s$, we refer here to $1\,Hz$ bandwidth.

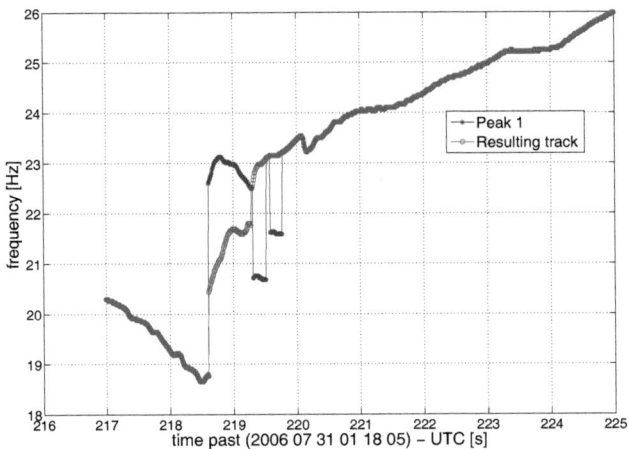

Figure 7.22: *VEX-VeRa occultation pass on DoY212-2006, orbit #101. Comparison of frequency vs. time profiles (data segment from the multipath region). The final profile from the "multi-track" routine (red) show smaller discontinuities than the profile obtained by the standard maximum-search algorithm does (blue). The "multi-track" profile needs to be smoothed in order to can feed the down-conversion stage of the following iteration step.*

which connects all first maxima detected in each spectrum, one per spectrum, one which connects all second maxima detected in each spectrum, one per spectrum, and so on. From all the available profiles, the routine must build a final profile by selecting a unique value of the frequency for each time instant. The first value will be the first entry of the profile which corresponds to the absolute maximum, since the routine starts to scan the spectra in a region where the the value of the SNR is still large enough to detect the signal with a tiny error probability. In each of the following step the routine scans all frequency entries which correspond to the time $t_n + 1$ and selects the frequency value $f_n + 1$ which is next to the previous one (f_n), regardless to which track it belongs (second maximum, third maximum and so on). The implicit assumption here is that signal coming from low altitudes in the atmosphere is quite stationary, as the antenna beam remains at constant height for more than the half duration of the whole pass between the engress- and the egress phases, so that it is unlikely that the signal sweeps more that one frequency bin from spectrum to spectrum.

Figure 7.23 compares the results of this process to the frequency profile obtained by the standard maximum-search routine.

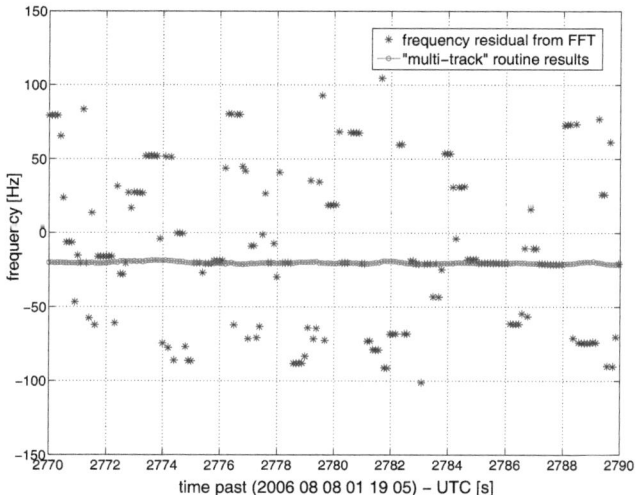

Figure 7.23: *VEX-VeRa occultation pass on DoY220-2006, orbit #109. Comparison of frequency vs. time profiles obtained by the standard maximum-search algorithm (blue) and by the "multi-track" routine (red).*

It is important to stress that the frequency profile obtained at each processing step is not the final profile. It rather represents the mixer signal of the down-conversion stage of the following processing step. This implies that it is possible to verify whether the "multi-track" routine has locked onto the signal or onto the noise. If the down-conversion signal were wrong, the mix-process would drive the signal towards the edge of the bandwidth, instead of centering its frequency content toward the center of the spectrum. This would be evident in the next processing stage, because of the increased observation time. In such cases, the multi-track routine may not be applied. In the praxis, since the signal attenuation increases with penetration depth, that is, with elapsed time in the retrieved frequency profile, one has to find the limit (time instant) at which the routine does not work any more (at ingress), and the time point from which the routine starts to deliver reliable results again (egress phase).

As the integration time is progressively increased during the iterative processing, the limit $SNR = 0\,dB$ is reached at each iteration step for a lower signal intensity w.r.t. the preceding stage (see figs. 7.10 to 7.13). This implies that the probing of the atmosphere can be progressively extended towards low altitudes. Typical values for minimum probed heights H are $H = 40\,km$ at beginning of the iterations down to $H \leqslant 35\,km$ at the end of the processing. Comparison between figs. 7.24 and 7.5 shows the noise reduction in the time intervals $600\,s$ to $1000\,s$ (ingress) and $2750\,s$ to $2900\,s$ (egress) achieved by means of the "multi-track" routine. This allowed an extension of the minimum probed height close to the limit of the height of super-refraction ($\sim 33\,km$), as it possible to see in fig. 7.25, which shows a comparison between the temperature profiles computed from both CL- and OL occultation data from the VEX-VeRa occultation pass on DoY220-2006. The figure shows a good agreement of the profiles between $80\,km$ and $45\,km$ altitude (height above ground, with assumed Venus radius $R_V = 6051.8\,km$), limit at which the CL recording stops. In the region comprised between $100\,km$ and $80\,km$ small discrepancies could be due to slight different boundary conditions and base-line fitting (see par. 8.2). Effects of multipath propagation affecting both temperature profiles in the region 55-60 km are considered in par. 7.2.3.

7.2.2 Processing techniques for multipath-affected signals

Effects of multipath propagation on occultation data have been extensively observed in Earth's atmosphere studies, such as the GPS/MET program ([30], [31], [91]), and the MIR/GEO experiments([53]). Multipath propagation was also observed in the Jupiter's ionosphere by the solar system exploration missions *Voyager 2* ([51]) and *Galileo* ([52]) and also in the atmosphere of Uranus by Voyager 2 ([63]).

This circumstance has lead to the investigation of different methods which could allow solving the ambiguity arising in the Doppler profiles of multipath-affected signals. Some of these approaches have proven to be effective and are now well established, whereas others innovative have been proposed recently. Hereafter a brief presentation of the two most commonly used categories of processing techniques for multipath-affected data is given: the *Radio-Holographic Methods*, and the methods based on the *Diffraction Theory*.[13]

---

[13]For a more comprehensive survey on the methods and algorithms which have been applied so

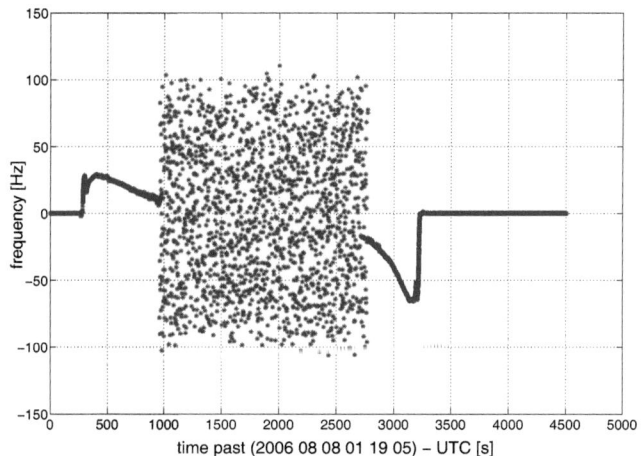

Figure 7.24: *VEX-VeRa occultation pass on DoY220-2006, orbit #109. Frequency vs. time profile, after the first iteration step, obtained by means of the "multi-track" routine, applied to data segments affected by multipath propagation ($t = 300$ to $t = 350$, and $t = 3180$ to $t = 3210$) and to data segments characterized by $SNR \sim 0$ ($t = 800$ to $t = 1000$, and $t = 2500$ to $t = 2900$). Comparison of this profile with fig. 7.5 reveals the noise reduction in the time intervals $600\,s$ to $1000\,s$ (ingress) and $2750\,s$ to $2900\,s$ (egress) achieved by means of the "multi-track" routine.*

Radio-Holographic Methods: the "Sliding-Spectra"-, or "Radio-Optic" Technique

The *Radio-Optic* technique (RO), or *Sliding-Spectra* technique (SS), which belongs to the category of the radio-holographic techniques ([75]), is based on complex spectral analysis of the time series representing the received signal (i.e., Fourier analysis), executed in *sliding mode*. Here, the extension of the time window, or *aperture size*, T corresponds to the Fresnel zone ([34]). The "quasi-stationarity" of the analyzed signal, needed for this kind of approach, is attained by down-converting the received signal by a model phase, which ideally "stops" the time-variability of the signal phase, or, at least, significantly narrows the signal dynamics, that is, the signal bandwidth ([69], [63]). The RO technique is -to a great extent- the technique which the OL data analysis software is based on, difference being that the spectral analysis

---

far to the processing of multipath-affected signals see [69].

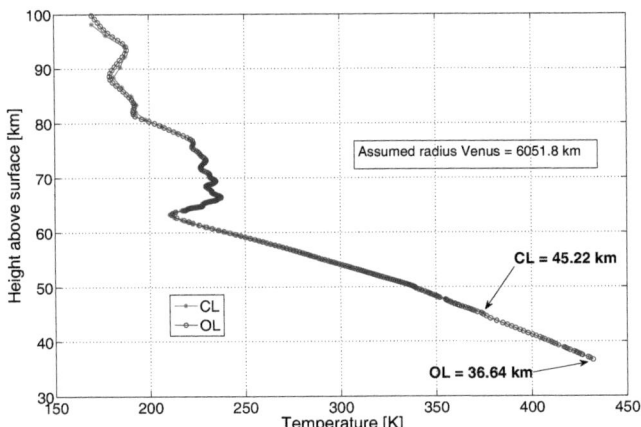

Figure 7.25: *VEX-VeRa occultation pass on DoY220-2006, orbit #109. Temperature profile. Comparison OL- vs. CL data. Implementation of the "multi-track" routine allowed signal recovery down to a minimum probed height close to the limit of the height of super-refraction (∼ 33 km). Data not corrected for multipath propagation effects.* ([95]).

implemented in the OL SW is applied iteratively and has only the purpose of slowing down and reposition the received signal in the center of the spectral window, while the conclusive analysis is conducted in the time domain, as mentioned in par. 7.1.

In the RO approach, each spectral maximum of each sliding spectrum is associated to a ray. The technique still works when more than one spectral peak is present, as in multipath events, as long as it is possible to distinguish peaks and associate a single ray to each one of them. The limit is reached in sub-caustic zones, where "false" maxima arises (see [33]; we interpret the *"false" maxima* as the interferences which occurs in the FFT when the simultaneously received signals get closer in spectrum). In order to overcome this limitation, different techniques have been proposed, which extends the capabilities of the RO method, some of which are briefly mentioned hereafter. Also the strategy we implemented in the OL software for the analysis of multipath-affected data is an enhancement of the RO method, achieved by means of the "Wigner-Ville Transform" (see par. 7.2.3).

A possibility to enhance the RO analysis has been proposed by Gorbunov ([33], [32]), who makes use of a "hybrid" method which complements the spectral analysis by means of the "Back-Propagation" technique (see after).

In [91] advantage is taken from expressing the amplitude and phase of the received signal as function of space, rather than time. In this way, instead of time windows, sliding spacial apertures are considered along the observational trajectory z (see fig. 7.26). The received complex signal $u(z) = a(z)\,exp[j\phi(z)]$ is decomposed in its spacial harmonics by means of spacial Fourier analysis; since harmonics are associated with rays arriving at each aperture, it is possible to relate a couple bending angle-impact parameter (ϵ_j, ρ_j) to the coefficient c_j of the correspondent Fourier harmonic. The final profile of the bending angle, $\epsilon(\rho)$ is obtained as sliding window averaging of all couples ϵ_j and ρ_j weighted by the relevant $|c_j|$. On one side, this approach has the significant advantage to allow for an automated (without supervision) computation of the bending angle, since it considers the whole spectrum for the weighted average, without identification and selection of local spectral maxima. On the other side, it presents the draw-back of the smearing-out of the retrieved profiles, due to the effects of the sub-caustic zones (false maxima entering the weighted average, [33]).

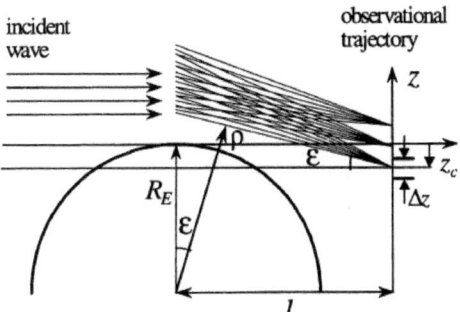

Figure 7.26: *Geometry considered for the computation of the bending angle ϵ and impact parameter ρ by means of the SS technique in the space domain. The spacial Fourier transform of the received complex signal is considered within small apertures Δz centered at z_c, sliding along the observational trajectory z. Single spacial harmonics are associated with rays arriving at each aperture. (Adapted from [91]).*

Interesting the analogy between high-resolution imaging radar and "focused synthetic aperture" radio-holography proposed by Igarashi *et al.* ([53]), who interprets the subtraction of a model phase from the received signal as a "focusing" operation, which increases the resolution of the angular spectrum of the received field (sum of the waves propagating at different angles). In the processing of SAR (Synthetic Aperture Radar) signals, the contributions of the elementary scatterers illuminated by the antenna beam are separated by means of a "matched filter" (MF), computed on the base of the orbital parameters of the measurement. The MF aligns and coherently sums up the phase-corrected phasors which correspond to the single scatterers. This is analogous to the subtraction of a model phase from the retrieved RO spectra. In fact, this operation makes the signal "stationary", thus allowing the extension of the integration (observation) time in the FFT (which corresponds to a coherent summation), with overall effect of improving the resolution (see par 7.1).

Methods based on Diffraction Theory: "Multiple-Phase-Screen" and "Back-Propagation"

These techniques are based on the *Helmholtz-Kirchhoff integral theorem* from classic electrodynamics, which expresses the amplitude and phase of an electromagnetic wave at a given point in terms of the integral of the distribution of amplitude and phase of the wave over a closed surface surrounding the point ([69], [29]). The theorem considers the amplitude and phase associated with either the electric- or the magnetic component of the wave (phasor), which is, therefore, regarded as a complex *scalar* function of the position ([29]). The theorem -valid when the scale of the radiating surface is very much larger than the considered wavelength- is at the base of the *Rayleigh-Sommerfeld* and *Fresnel-Kirchhoff scalar diffraction theories*. These theories enable both the forward-propagation technique, called *Multiple-Phase-Screen* (MPS), and the *Backward-Propagation* (BP) technique ([69], [30]).

The MPS technique is used to solve the wave propagation problem when simulating radio occultation observations from model atmospheres ([57], [91], [50], [31]). It is based on a *phase screen model* (see fig. 7.27), which is briefly described in the following. The transmitter, on the left-hand side of fig. 7.27 (not shown) is supposed infinitely far in space. This is equivalent to considering plane waves entering the atmosphere.[14] The three-dimensional (3-D) atmospheric refractivity is

---

[14]The large distance to the transmitter, as compared to the vertical scale of the atmosphere, allows this approximation.

replaced by a number of parallel phase screens normal to the direction of the incident wave and the signal is "observed" on the plane at a distance l_0 from the limb. Approximation of the phase screen is applicable when the deviation of rays from the straight lines inside a refractive medium is smaller than the smallest scale of the refractivity irregularities that are significant for a given propagation problem. Thus, the approximation of the whole atmosphere by a single phase screen is not applicable. Instead, multiple phase screens have to be used, so that the aforementioned condition is satisfied between the adjacent screens.

The excess phase (phase path) s, assigned to a point on a phase screen, is equal to refractivity $N(h)$ (being h the radial distance from the ground) integrated between the adjacent screens along the straight line normal to the screens. It is assumed that the refractivity is spherically symmetric with the vertical profile. Since the Fresnel zone on a phase screen is much smaller than the radius of the considered planet, instead of the 3-D problem a 2-D problem in Cartesian coordinates (y, z) is considered, as shown. For a small enough distance Δl between the adjacent screens it is sufficient to calculate the excess phase profile $s(z)$ approximately, by multipiying Δl by the refractivity profile $N(h)$, taking into account its shift z_{sh} and tilt θ with respect to the screen at a distance l from the limb as shown in fig. 7.27:

$$s(l, z) = 10^{-6} N \frac{z - z_{sh}}{\cos\theta} \Delta l , \qquad (7.23)$$

where $z_{sh} = R_E(\cos\theta - 1)$ and $\cos\theta = (1 - l^2/R_E^2)^{1/2}$ ([91]).

The back-propagation technique is the dual of the MPS technique, with the direction of propagation inverted. Starting by the field as observed at the reception point, the field at a point in space located behind the receiver toward the transmitter is calculated by numerically propagating back the received wave ([57], [30], [50], [52], [51]).

The principle explanation of this method presented hereafter follows the formulation of [52]. First, considering again the simplified occultation geometry of fig. 7.27, the field recorded along the observational trajectory as a function of space is

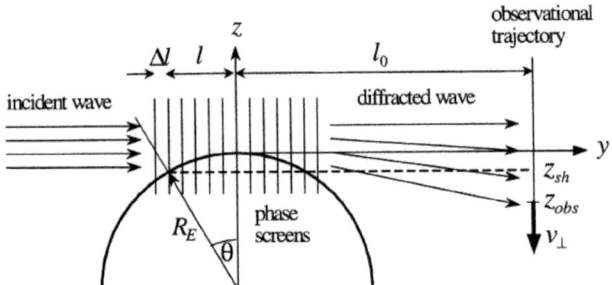

Figure 7.27: *Layout of the wave propagation problem in the atmosphere with the use of the multiple phase screens ([91]).*

represented as a spectrum of plane waves by means of spacial Fourier transformation:

$$U(k_z, y = l_0) = \int_{-\infty}^{\infty} u(z, y = l_0) e^{-jk_z z} \, dz \ , \tag{7.24}$$

where k_z is the wave number in the z direction. The corresponding wave number in the y direction is:

$$k_y = \sqrt{k^2 - k_z^2} \ , \tag{7.25}$$

where $k = 2\pi/\lambda$. ($k_x = 0$ due to the assumed spherical symmetry).

Each Fourier component corresponds to a plane, electromagnetic wave propagating in the direction specified by the wave number (k_y, k_z). The angular spectrum along an arbitrary line $y = l'_0$ is obtained by applying the appropriate phase shift to each component of the spectrum at $y = l_0$. For propagation in vacuum, it follows that:

$$U(k_z, l'_0) = U(k_z, l_0) \exp[jk_y (l'_0 - l_0)] \ , \tag{7.26}$$

The field at this new location is obtained by combining the phase shifted Fourier

176

components (i.e., through inverse Fourier transform):

$$u(z, l'_0) = \frac{1}{2\pi} \int_{-\infty}^{\infty} U(k_z, l'_0) \, e^{jk_z z} \, dk_z \;, \tag{7.27}$$

Equations 7.24 to 7.27 are equivalent to a *Huygens-Fresnel* diffraction integral ([29]).

As shown in fig. 7.28, the BP technique is exploited to correct the chronological sequence of phase- and amplitude values of the received signal, which was altered by multipath events. The received field is back-propagated, or *mapped*, into another surface in order to obtain a new virtual time series of phase- and amplitude values which has no multi-valued Doppler points ([69]).

This concept is illustrated in fig. 7.29, which shows how the back-propagation method is based upon the *Helmholtz-Kirchhoff* theorem. Again, the case of a GPS receiver installed in a LEO satellite orbiting the Earth is considered. The field received by the LEO along the curve S is multipath-affected. Instead, an ideal receiver traveling the line containing the point r would not experience multipath reception: the values of the received phase along this line would provide monotonic profiles of the *bending angle* α vs. the *impact parameter* p. Due to the Helmholtz-Kirchhoff theorem, the field received along the curve S is *equivalent* to the field along the line containing the point r.[15] According to the theorem, each phasor considered along the line containing r is calculated by integrating all infinitesimal contributions along the curve S. The measurements of the complex field along S constitute the boundary conditions for the Helmholtz-Kirchhoff theorem. The back-propagated field is calculated using the diffractive integral ([32]):

$$u(\mathbf{r}) = \left(\frac{k}{2\pi}\right)^{1/2} \int_S u_0(\mathbf{y}) \times \cos\varphi_{\mathbf{ry}} \, \frac{\exp(-jk|\mathbf{r}-\mathbf{y}|+j\pi/4)}{|\mathbf{r}-\mathbf{y}|^{1/2}} \, dS_{\mathbf{y}} \;, \tag{7.28}$$

where $\varphi_{\mathbf{ry}}$ is the angle between vector $\mathbf{r}-\mathbf{y}$ and normal $n_{\mathbf{y}}$ to curve S at current

---

[15] As in the case of the MPS technique, due to the assumed spherical symmetry of the refractivity, the problem is threated in a 2-D geometry. Therefore, the integration over a surface enclosing the point r reduces to the integral along the curve S, since the field is null over the complete surface with exception of the curve S.

integration point **y**, and $u_0(\mathbf{y})$ is the boundary condition. Curve S is the LEO trajectory.

It has to be noticed that although the back propagation is performed as in vacuum, the back propagation plane can (and must) be positioned inside the atmosphere, and the back-propagated field is thus not the real wave field at the location of the back propagation plane ([32]). But since back propagation in a vacuum is equivalent to straight line continuation of rays (i.e., it preserves their impact parameters), even though the back propagated field is not equal to the true electromagnetic field in the atmosphere, it provides the same bending angle as a function of the impact parameter ([91]).

A problem presented by the BP technique is the choice of the location of the back propagation plane, which should be set in a single-ray area in order to eliminate the effects of multipath. Therefore, the ideal position depends on the caustic structure, which is a priori unknown ([33]).

In the considered geometry, as for the case of MPS, the transmitter (GPS satellite) was considered at infinite distance from the probed atmosphere in order to can use the plane waves approximation. The same geometry is involved in the investigation of extra-terrestrial planetary atmospheres, with the only difference that sender and receiver are swapped: the S/C transmits from the vicinity of the planet, whereas the receiver (i.e., the G/S) is considered infinitely remote. Duality holds, so that the discussed techniques apply as well.

An interesting observation is that the back-propagation method reduces the effect of multipath propagation because it increases the *resolution power* of the observation. Taking the example of a S/C sounding an extra-terrestrial atmosphere, mapping the transmitted field in the back-plane corresponds to reduce the distance d between the transmitter and the atmosphere. This reduces the vertical extension of the first Fresnel zone, $2\sqrt{\lambda d}$, λ being the considered wavelength (see par.5.2.4). As consequence, the improved vertical resolution of the measurement allows to better resolve diffractive structures, such as small-scale perturbations of the refractivity, thus mitigating multipath propagation effects ([50], [57]). Han and Tyler ([50]) consider the case of *Mars Global Surveyor*, for which a given S/C distance from the planet of $1750\,km$ corresponds to a diameter of the first Fresnel zone of about $500\,m$ (about twenty times larger than the value obtained for low sounding altitudes at

Venus for the same distance S/C planet, due to the defocusing effect of the thick Venus' atmosphere, see par. 5.2.4). By application of the BP technique to simulated data, they could resolve a $40\,m$ small-scale refractivity perturbation of magnitude 10^{-7}, thus increasing the resolution of one magnitude order over the limit imposed by the Fresnel zone.

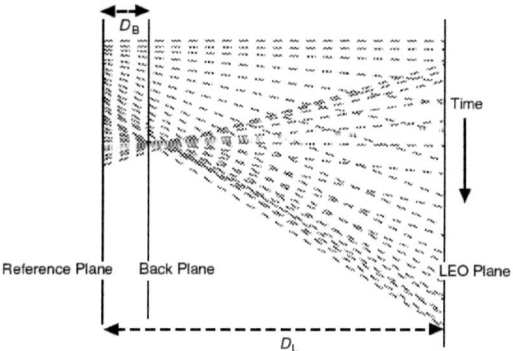

Figure 7.28: *Back propagation geometry. In order to reduce the effects of multipath, the field measured in the LEO plane (LP) is back propagated to a back plane (BP) set to a distance D_B from a reference plane (RP) in which the rays have not yet crossed each other. The distance D_L separates the RP from the LP. The choice of the BP position (distance D_B) is not trivial, since the BP should be set in a single-ray area, but the caustic structure is a priori unknown ([33]). Adapted from ([69]).*

7.2.3 The Wigner-Ville Transform

As well expressed in [69]: *"what this multipath problem[...]needs is a transformation that converts the multi-valued time series of observations (in Doppler or bending angle) into a single-valued series."* When looking at multipath spectra, as in figs. 7.18 and 7.19, one is confronted with the embarrassing question, which frequency value is to be attributed to those time instants at which multiple spectral peaks are detected. An important step toward the answering of this question is to notice that each spectral speak, that is, each carrier signal which shows itself in the Fourier spectrum corresponds to a different "ray". This implies that the signals associated with different rays come from different regions of the transected atmosphere.

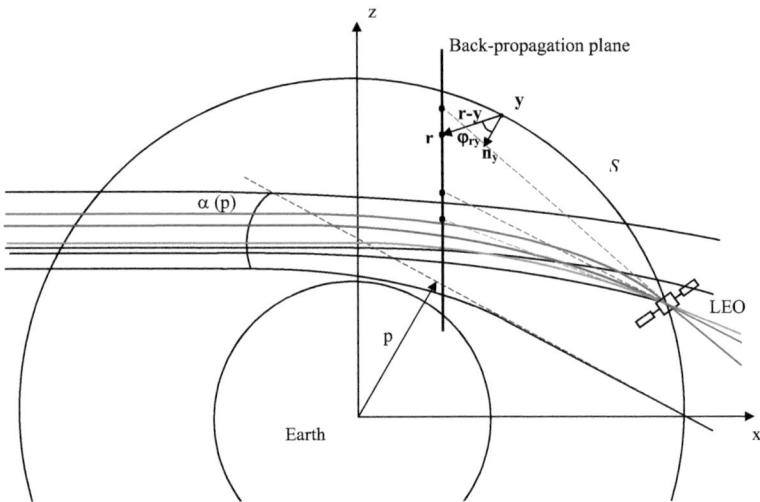

Figure 7.29: *Multipath propagation geometry and back-propagation (for the example of a LEO satellite). Non-monotonous vertical profile of the bending angle $\alpha(p)$ determines multipath propagation (rays which cause interference at LEO are colored). Back-propagation of the wave field (straight-line continuation of rays, shown as dashed lines, in the same colors of the respective rays) restores the correct vertical sequence of the rays. Back-propagated field at point $\mathbf{r}$ is calculated via diffractive integral as a superposition of infinitesimal fields from the points of curve S. Each point $\mathbf{y}$ can be represented as a transmitter with intensity $u_0(\mathbf{y})$ and direction diagram $\cos(\varphi_{ry})$. (Adapted from [32]). Even though the back propagated field is not equal to the true electromagnetic field in the atmosphere, it provides the same bending angle as a function of the impact parameter ([91]).*

Keeping in mind that the attitude maneuvers executed by the spacecraft during the occultation event in order to track the virtual image of the Earth through the Venus atmosphere were based on an atmospheric model which has spherical symmetry and constant refractivity gradient, it is possible to convince oneself that each of the received rays crosses the atmosphere at different altitudes, or, equivalently, each atmospheric height is scanned only once ([79], [51]), so that for each height there is a unique value of the Doppler shift, a unique value of the bending angle and a unique value of the impact parameter ([98]). This opens the way to the idea that what is needed in order to correctly interpret the frequency of the received signals is to bring the received Doppler shifts in a order which is not chronological, but "spatial", that is, one has no longer to interpret the frequency as a function of the time, but as a function of the height. This reverses the problem: if one were to order the series of the Doppler value along a continuous line which tracks the spectral peak throughout the multipath event (see fig. 7.30), the time order would follow. This can be achieved by looking at the sequence of the spectral maxima as a continuous series, each peak occurring at a given frequency.

At this point, the issue of major concern is how to obtain a suitable spectral representation of the data which allows recovering the value of the *instantaneous frequency* of each of the received signals at each given time instant. In fact, as already mentioned, the FFT-based analysis presents the draw-back of conflicting requirements between time- and frequency resolution. On one hand, if one can find an appropriate time interval which is short enough for the component signals to be considered stationary, the relevant frequency resolution is not good enough to resolve close spectral peaks. On the other hand, the frequency resolution cell (frequency bin) cannot be arbitrary reduced, as the consequent increase of the integration time would let the signal frequency "run" over the observation window, thus smearing the frequency information over several frequency bins. What is needed is a convenient *time-frequency representation* of the received signal(s).

As simultaneously and independently reported by [35] and [80], the *Wigner Distribution*, or *Wigner-Ville Transform*[16] has proven to be a suitable analysis tool for

---

[16]The Wiener Distribution (WD) is named after Eugene Wigner, the Hungarian American physicist who first introduced it in quantum mechanics ([103]). This operator is also known as Wigner-Ville Distribution, as it was later re-derived by J. Ville in 1948 as a quadratic (in signal) representation of the local time-frequency energy of a signal ([102]).

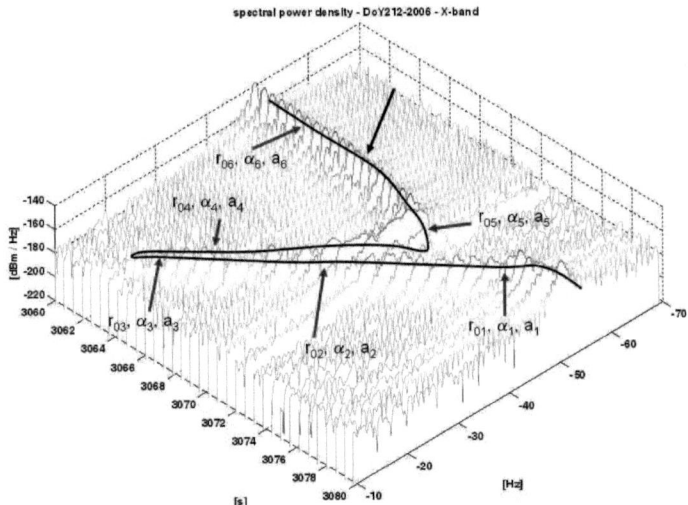

Figure 7.30: *VEX-VeRa occultation pass on DoY212-2006, orbit #101. Waterfall diagram showing the time sequence of power density spectra of received signal in the multipath region during the egress phase as in fig.7.19. The black line connecting the spectral peaks suggests a possible new arrangement of the frequency points, which is no longer chronological but "spacial". At each frequency point $\mathbf{f_i}$ corresponds a unique value of the atmospheric height $\mathbf{r_i}$, bending angle $\boldsymbol{\alpha_i}$, and impact parameter $\mathbf{a_i}$ ([80]).*

solving ambiguity in multipath regions[17]. This integral operator is a *bilinear map* $\mathbf{V} \times \mathbf{V} \to \mathbf{F}$, where $\mathbf{V}$ is a vector space (i.e. the vector space of the time signals[18]) and $\mathbf{F}$ is a scalar field (i.e., $\mathbb{R}$):

$$W_{f,g}(t,\omega) = \int_{-\infty}^{\infty} e^{-j\omega\tau} f\left(t+\frac{\tau}{2}\right) g^*\left(t-\frac{\tau}{2}\right) d\tau \qquad (7.29)$$

where the functions $f(t)$ and $g(t)$ are assumed to be (in general) complex time-signals and the asterisk denotes the complex conjugated. The quantity $W_{f,g}(t,\omega)$ is

---

[17]A brief presentation of the WD and of some of its properties which are of particular relevance for signal processing are given in App. G.
[18]It is well known that exponential functions of the kind $e^{j\omega_i t}$ constitute an orthonormal base upon which the set of the time-signals has a space vector structure.

a real function of the two variables t and ω, which have the dimension $[s]$ and $[s^{-1}]$, respectively, that is, the WD is a function of time and (angular) frequency.

Considering $g(t) = f(t)$ leads to:

$$W_f(t,\omega) = \int_{-\infty}^{\infty} e^{-j\omega\tau} f\left(t + \frac{\tau}{2}\right) f^*\left(t - \frac{\tau}{2}\right) d\tau \qquad (7.30)$$

The distribution $W_f(t,\omega)$ is capable of rendering the spectral contents of the signal $f(t)$ at each given time instant t_i with a frequency resolution which is independent of the time.

The WD was tested on the data set whose spectrogram is shown in fig. 7.19, obtained from the VEX-VeRa occultation pass on DoY212-2006, with the purpose of recovering a continuous frequency trace through the spectral peaks, as shown by the black line in fig. 7.30. The result is shown in fig. 7.31. Even if surrounded by a noisy background (due to the so-called "cross terms" of the WD), a clear frequency profile is visible. It is marked by a continuous blue line in fig. 7.32, obtained by tracking the maxima of the distribution along the frequency axis.

The recovery and the proper alignment of the frequency values made it possible to calculate an unambiguous profile of the bending angle vs. time, which, in turns, allowed correct calculation of refractivity and temperature even in the multipath region. Figure 7.33 shows a comparison of temperature profiles obtained by means of OL- and CL data from the multipath region in the "egress" phase of the same example pass (DoY212-2006). Figures 7.34 and 7.35 show a comparison of results from CL data and OL data with WD application ([95]).

When converted to physical quantities, the effects of the implementation of the WD are conspicuous. Comparing temperature profiles obtained respectively with- and without implementation of WD (see figs. 7.33 and 7.35), we notice that the temperature inversion in the Venus atmosphere at $\sim 58\,km$ becomes more pronounced . It is beyond the scope of this work to discuss the physical consequences but it is very likely that this will have an effect on wind fields and stability parameters in the atmosphere ([49]).

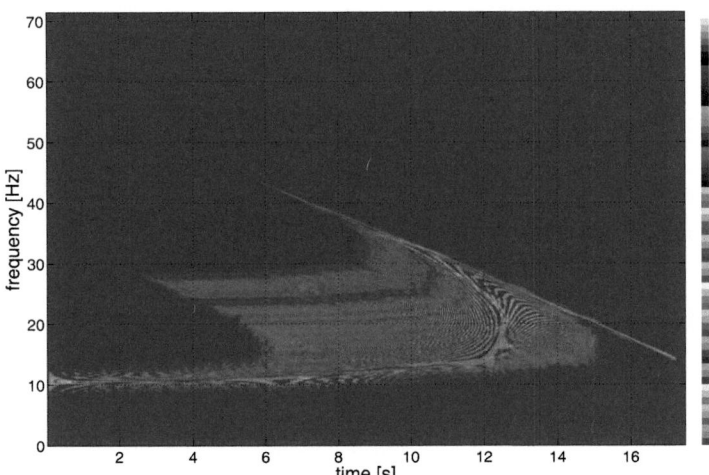

Figure 7.31: *VEX-VeRa occultation pass on DoY212-2006, orbit #101. Wigner-Ville Transform of received signal in the multipath region during the egress phase. A clear visible curve emerges from background noise which connects the function maxima along the frequency variable ([80]).*

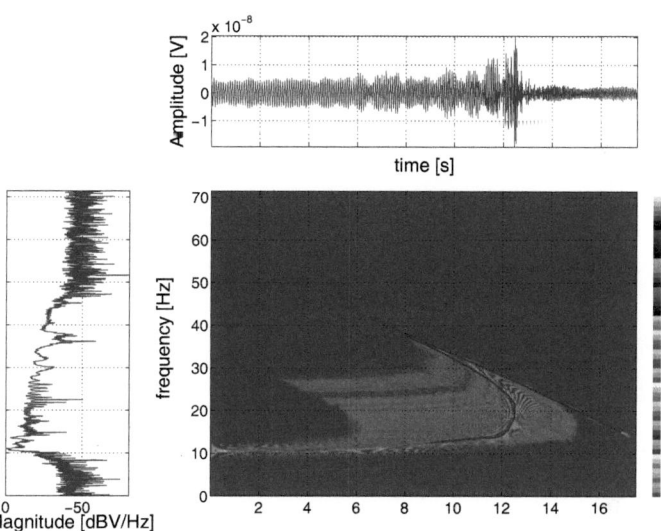

Figure 7.32: *VEX-VeRa occultation pass on DoY212-2006, orbit #101. Wigner-Ville Transform of received signal in the multipath region during the egress phase. The same as in fig. 7.31 with explicitly connected maxima (blue line). On the top panel the analyzed samples in the time domain. The time signal show a "beating" which typically happens when two carrier signals are received simultaneously. The frequency of the envelope of the beating correspond to the frequency separation of the two components. On the left-hand-side panel the FFT of the whole data set ([80]).*

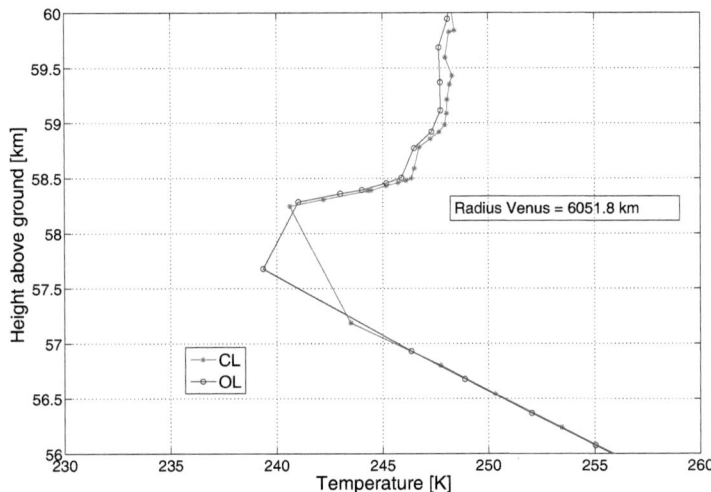

Figure 7.33: *VEX-VeRa occultation pass on DoY212-2006, orbit #101. Temperature profile. Comparison OL- vs. CL data. Particular of the region of multipath. The "multi-track" routine and the repeated Low-Pass filtering, applied at each processing stage with a smaller value of the cut-off frequency, have produced a smooth temperature profile, which, however is not completely representative of that region of the atmosphere, as part of the incoming rays has been filtered out and the chronological sequence of the processed rays does not correspond to a monotonic height profile. The CL data show a similar result, which was obtained by averaging the phase values obtained from the receiver PLL ([95]).*

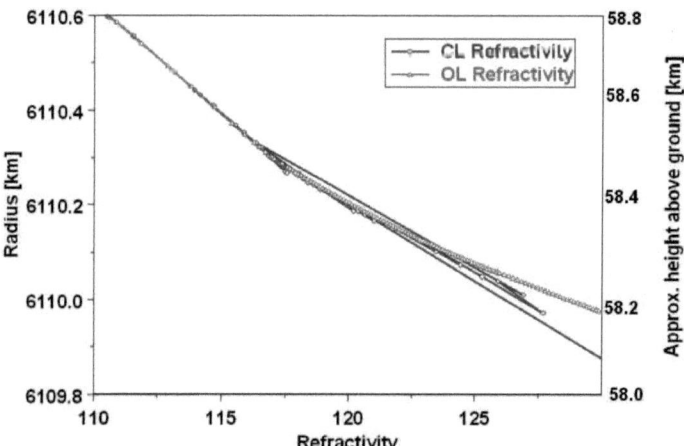

Figure 7.34: *VEX-VeRa occultation pass on DoY212-2006, orbit #101. Refractivity profile. Comparison OL- vs. CL data. Particular of the region of multipath. CL data present themselves as a non-monotonic profile when connecting the single data point in a chronological order. The OL refractivity curve was derived by a frequency profile whose calculation was supported by the WD in the multipath region ([95]).*

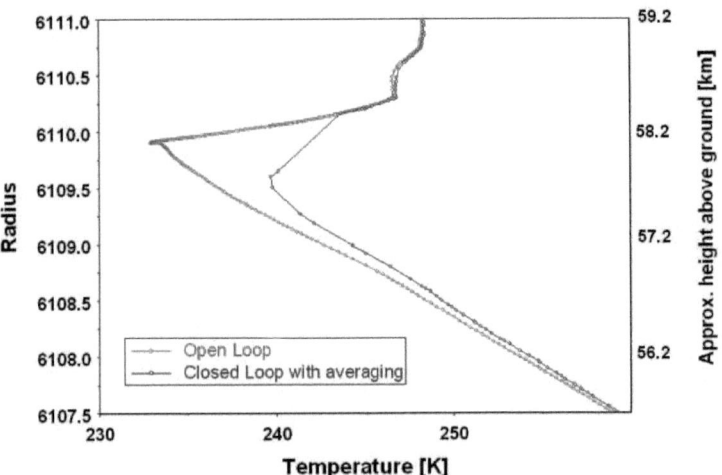

Figure 7.35: *VEX-VeRa occultation pass on DoY212-2006, orbit #101. Temperature profile from the multipath region. Comparison OL- vs. CL data (same data-set as in 7.34). The necessity of averaging the data points in the CL profile leads to an inaccurate temperature profiles, whereas the OL could recover the right spatial sequence of the incoming rays by means of the WD distribution ([95]).*

Chapter 8

Error Analysis

Experimental observations are affected by different kinds of errors, originating from several sources. In order to evaluate the results obtained from the reduction of the Occultation *Open Loop* raw data, the possible error sources and their effect on the processed data are considered in this chapter.

Errors are divided into two general categories:

- *statistical error* (or *random error*). In this category fall:

    - receiver thermal noise [1] ;
    - phase noise of the on-board oscillator (USO);
    - quantization noise;
    - clocking jitter at the ground-station;
    - fluctuations in the propagation path;

- *systematic error*. This category comprehend all "deterministic" errors, i.e. errors whose outcome is the same at each repetition of the experiment. (For this reasons, this errors are said to bring a *bias* in the measurement). Most common systematic errors in space missions are:

    - inaccuracy in the evaluation of the effects of the Earth ionosphere and troposphere;

---

[1] As " receiver thermal noise" is meant the thermal noise considered at the receiver input, given by the system temperature T_{Sys}, therefore including the antenna contribution, too (see App. B).

- mis-pointing of the S/C antenna;
- oscillator drift (aging);
- inaccuracy in the reconstructed orbit data and in the S/C attitude;
- execution time offset;
- computational errors due to the finite precision number representation of calculators;
- computational errors due to numeric implementation of continuous operations (i.e., integration and derivation);
- errors introduced at specific places by the algorithms for data evaluation (depend on the adopted algorithms);

8.1 Statistical Error

8.1.1 Estimation of the receiver thermal noise

The noise-induced uncertainties in the calculated frequency- and power profiles are expressed as a function of the Signal-to-Noise Ratio (SNR, see App. B), which is a time-varying quantity since the level of the received signal changes throughout the occultation pass.[2] In order to estimate the intensity of the received signal, the baseline noise power must be estimated and subtracted from the total received power.

The receiver thermal noise is assumed to be additive white Gaussian noise (AWGN), modeled as ergodic stochastic process on both *in-phase-* and *quadrature* channels. It is assumed that the processes are independent.

The estimation of the noise power is performed in the pre-processing step of the Open-Loop software (see Ch.7). The noise process is characterized as *locally* WSS (see par.6.3): its variance (noise power) is regarded as a function of time whose values are estimated over non-overlapping time intervals of $1s$. In order to reveal long-term variations of the noise floor, the time series of the obtained noise estimate is not averaged (see fig. 8.1).

---

[2]Small fluctuations of the receiver noise floor take place as well, but, since estimates of the noise power density observed during several occultation measurements reveal variations in the range of 0.05 dB, they are considered as not significant for the purpose of error analysis.

The estimate of the noise power is based on the periodogram method: for each considered segment of data, the DFT is calculated and all values in the spectrum (periodogram) are integrated (summed) over the frequency axis, excluding a $20\,kHz$ window centered around the signal peak, and divided by the number of the considered frequency bins. (Further division by the considered bandwidth yields the power spectral density). This operation corresponds to averaging *within* each periodogram (*frequency* averaging), as an alternative to averaging *different* periodograms (*time* averaging), which would lead to prolonging the assumed stationarity time interval and diminishing the time resolution.

As an example, fig. 8.1 shows the NNO receiver noise power spectral density as a function of time during the VEX-VeRa occultation pass on August, 8th, 2006 (orbit #109).

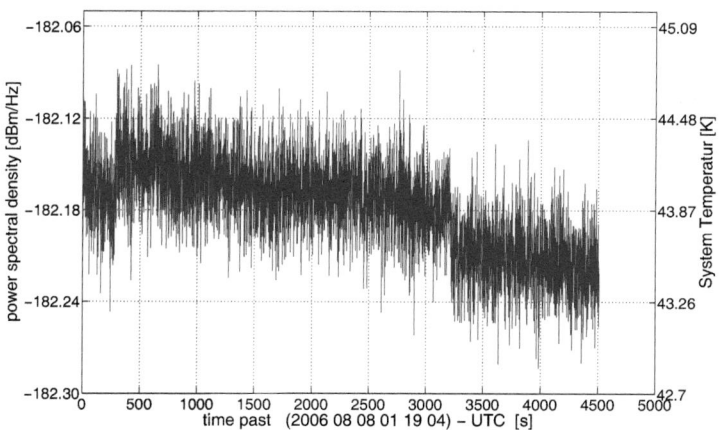

Figure 8.1: *Estimate of the receiver noise power density at the NNO G/S during the VEX-VeRa occultation pass on DoY220-2006, orbit #109. Each point was calculated by averaging the relevant periodogram in segments distant from the signal peak (excluded frequency interval: $\pm 10\,kHz$, centered at the spectral maximum). Transitions at $t = 286$, and $t = 3218$ are due to the switch of the amplification level in the AGC circuit (see after).*

In the following, we perform a consistency check of the obtained estimate. To this purpose, the calculated noise estimate is shown again in fig. 8.2 in linear units of *zeptowatts* (zW) pro Hertz, where $1zW = 10^{-21}W$.

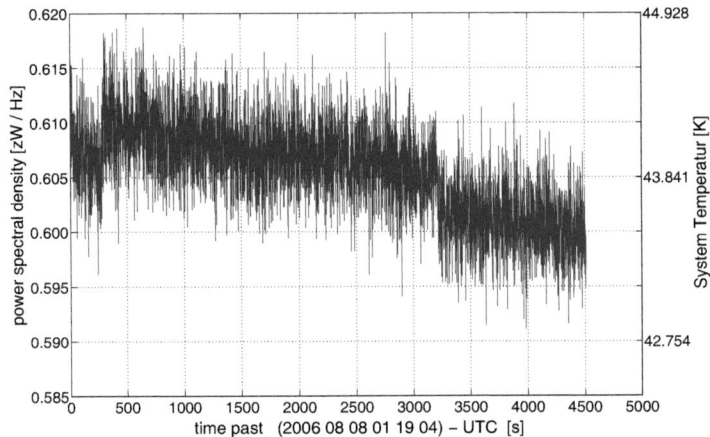

Figure 8.2: *The same as fig. 8.1 but in linear units ($1zW = 10^{-21}W$).*

The statistics of the periodogram of white noise, given in eqs. (6.100), and (6.102) are reported hereafter for convenience:

$$\mathscr{E}\{\hat{P}_{PER}(f)\} = P_{xx}(f) = N_0 , \qquad (8.1)$$

and

$$\sigma^2_{\hat{P}_{PER}}(f) \simeq P^2_{xx}(f) = N^2_0 , \qquad (8.2)$$

where $P_{xx}(f)$ represent the true power density spectrum, and N_0 is the noise power density (the true value of the parameter to be estimated). The average performed in each periodogram should reduce the variance of the estimate by a factor which equals the number N of the averaged points. In order to verify this, the statistics of the estimate are qualitatively assessed by means of fig. 8.2 and compared with the relevant theoretic values.

Since from eqs. (8.1), and (8.2) follows:

$$\mathscr{E}\{\hat{P}_{PER}(f)\} = \sigma_{\hat{P}_{PER}}(f) , \qquad (8.3)$$

the value of the standard deviation of each periodogram (standard deviation before averaging) can be obtained as the expected value of the points in fig. 8.2. The value $\hat{N}_0 = 0.608 \, zW/Hz$, obtained by looking at the graphic in the time interval comprised between $t = 1500$ and $t = 3000$, is assumed as expectation of the periodogram. Dividing this value by the square root of the number N of the points considered for the average gives the "theoretic" value of the standard deviation of the estimate:

$$std\{\hat{P}_{PER}(f)\} = \frac{\hat{N}_0}{\sqrt{N}} = \frac{0.608}{\sqrt{N}} \simeq 0.0025 , \qquad (8.4)$$

whit $N = 6 \times 10^4$. The obtained value shall be near to the value of the standard deviation calculated from the points on the graphic, which is the difference between a value at 1σ distance from the expected value and the expected value itself. Looking at the graphic in the same time interval, the values lying within $\pm 1\sigma$ distance from the expectation are considered to be comprised between $0.606 \, zW/Hz$ and $0.610 \, zW/Hz$; the standard deviation of the estimator is therefore computed as:

$$std_{gr}\{\hat{P}_{PER}(f)\} = 0.610 - 0.608 = 0.002 , \qquad (8.5)$$

which confirms the assumption (subscript "gr" stands for graphic, meaning that the standard deviation was obtained from a qualitative analysis of the graphic). The value $\hat{N}_0 = 0.608 \, zW/Hz$ ($\hat{N}_{0|dB} = -182.17 \, dBm/Hz$) is assumed as baseline noise power density N_0 for the calculation of the SNR_0 in the following paragraph.

At $t = 286$, and $t = 3218$, the AGC circuit switched the level of the amplification. The residual value of the two transitions after calibration are still to be seen in the graphic, since they are under the specified accuracy of the gain calibration curve ($0.1 \, dB$, [55]). Fig. 8.3 shows the noise curve after correction of the inaccuracy of the gain calibration curve. An offset of $0.02\,dB$ was added to the values in the segments which go from the beginning to $t = 287$, and from $t = 3219$ to the end, respectively.

After correction it is possible to asses a linear drift of the noise power density of $2 \times 10^{-6}\,zW/Hz/s$. The noise power density difference between beginning- and end of the measurement is $\Delta \hat{N}_0 = 9 \times 10^{-3}\,zW/Hz$ ($\sim 1.5\%$ of the mean value), which corresponds to a change in the system temperature (see (B.11), and (B.12)) of $0.65\,K$. This small amount could be attributed to different illumination conditions of the antenna during the tracking (i.e. sun peeking through a side-lobe of the antenna pattern, or being "seen" by the main lobe under an angle closer to the zenith (maximum antenna gain), or to the effects of the cooling of the receiver.

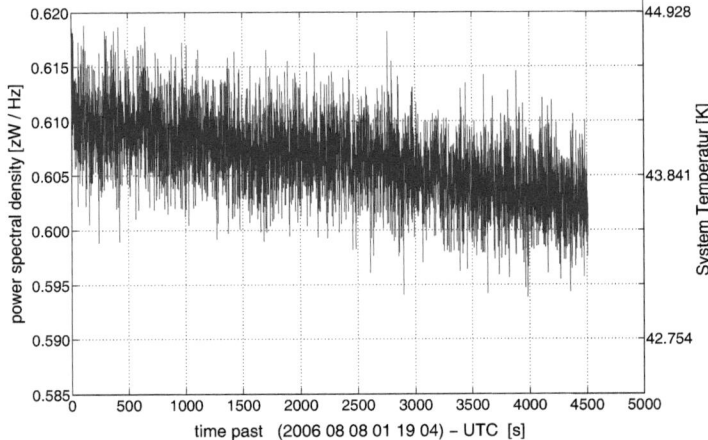

Figure 8.3: *The same as fig. 8.2 with correction of calibration inaccuracy in the segments which go from the beginning to $t = 286$ and from $t = 3219$ to the end, respectively.*

8.1.2 Uncertainties in the calculated frequency- and power profiles

Statistics

In order to assess the effect of the random error on the results, theoretic statistics applying for modeled signal and noise are first considered. The analysis is conducted for two cases: the case in which no signal is received (only noise), and the case of superimposed sinusoidal deterministic signal (noise plus signal). The theoretic statistics are then compared with the actual distribution of the data in order to validate the error analysis of the calculated frequency- and power profiles.

The receiver thermal noise considered at the output of an ideal narrow-band filter cascaded to the receiver and centered at the nominal carrier frequency ω_c can be represented as:

$$n(t) = x(t)\,cos(\omega_c t) - y(t)\,sin(\omega_c t) , \tag{8.6}$$

where $x(t)$ and $y(t)$ are zero-mean, normal stochastic processes which represent the base-band, or *video* components of the noise in the in-phase and quadrature channel of the receiver, respectively ([77]). They are uncorrelated (and therefore independent, since normally distributed), with jointly normal probability density function:

$$p(x,y) = p(x) \cdot p(y) = \frac{1}{2\pi\sigma_n^2} e^{-\frac{(x^2+y^2)}{2\sigma_n^2}} , \tag{8.7}$$

where σ_n^2 is the noise power given by the bandwidth of the considered filter (see eq. (B.10)). (It has to be noticed that $\sigma_n^2 = \sigma_x^2 = \sigma_y^2$, that is, the noise power of each channel coincides with the noise power at the output of the filter).

Expressing the (base-band) noise in polar coordinates yields:

$$\begin{cases} r(t) = \sqrt{x^2(t) + y^2(t)} \\ \phi(t) = \arctan\left[\dfrac{y(t)}{x(t)}\right] \end{cases} \qquad (8.8)$$

where $r(t)$ is the noise *amplitude* and $\phi(t)$ is the noise *phase*. It is possible to show that $r(t)$ follows a *Rayleigh* distribution, whereas $\phi(t)$ in uniformly distributed in the phase interval $\pm\pi$ ([77]). If a deterministic, harmonic signal of amplitude A at carrier frequency ω_c is superimposed to the noise, the amplitude and phase detected at the filter output become, respectively:[3]

$$\begin{cases} r(t) = \sqrt{[x(t) + A]^2 + y^2(t)} \\ \phi(t) = \arctan\left[\dfrac{y(t)}{x(t) + A}\right] \end{cases}, \qquad (8.9)$$

with *signal-to-noise* ratio at the filter output given by:

$$SNR = \frac{A^2}{2\sigma_n^2}, \qquad (8.10)$$

which determines the statistics of the quantities in the relationships (8.9). It is possible to show ([77]) that, for $A \gg \sigma_n$, the amplitude tends to be normally distributed around the expected value A with variance σ_n^2, whereas for $A = 0$ it follows the Rayleigh distribution, as expected. Therefore, for high values of the SNR, the uncertainty in the detected signal amplitude coincides with the standard deviation of the noise:

$$\sigma_A = \sigma_n, \qquad (8.11)$$

---

[3]In order to simplify the formalism, the signal is considered to have zero phase offset at the start time, so that it is present in only one channel; this assumptions has no conceptual relevance, as the phase offset of the considered signal depends on the considered start time on the time abscissa.

whereas, the relative error is:

$$\varepsilon_A = \frac{\sigma_n}{A} = \frac{1}{\sqrt{2}} \cdot \frac{1}{\sqrt{SNR}} \qquad (8.12)$$

Under the same condition of high SNR, also assuming small phase increments of the phase detector output ($|\Delta\phi| < 5°$) [4], the phase follows a normal distribution with expectation zero and variance $\sigma_\phi^2 = \sigma_n^2/A^2$ ([77]), that is, the variance of the phase error $\Delta\phi$ is inversely proportional to the SNR: [5]

$$\sigma_\phi^2 = \frac{1}{2\,SNR} \qquad (8.13)$$

The variance of the frequency error can be derived from the variance of the phase error:

$$\sigma_f^2 = \left(\frac{\sigma_\phi}{2\pi\Delta t}\right)^2, \qquad (8.14)$$

where Δt is the sampling time. Therefore, the uncertainty in the evaluation of the signal frequency is:

$$\sigma_f = \frac{1}{2\pi\Delta t} \cdot \frac{1}{\sqrt{2}} \cdot \frac{1}{\sqrt{SNR}} \qquad (8.15)$$

The signal *peak* power is defined as $P = A^2$. In order to evaluate the inaccuracy of the retrieved times profile of the received signal power, the first two moments of the random variable P are considered. To this purpose, we recall that, for the case of high SNR ($A >> \sigma_n$), the detected amplitude is a r.v. normally distributed around the signal amplitude A with variance σ_n^2. Writing the peak power P as the

---

[4]The phase detector evaluates the *incremental* phase of the signal by computing the difference between the phase of the input signal and the phase of the local oscillator, tuned at the same frequency of the signal. For a non-modulated, harmonic input signal, in absence of noise the phase difference is a constant.

[5]The same results is obtained for the phase error of a PLL, see [104].

square of the sum of the signal amplitude and the noise amplitude n (see eq. (8.6)):

$$P = (A+n)^2 , \qquad (8.16)$$

it is possible to calculate expected value and variance of P as:

$$\bar{P} = E\{(A+n)^2\} = A^2 + \sigma_n^2 , \qquad (8.17)$$

and

$$\sigma_P^2 = E\{(P-\bar{P})^2\} = E\{(A+n)^2 - (A^2+\sigma_n^2)\} = 2\sigma_n^2 (2A^2 + \sigma_n^2) , \qquad (8.18)$$

respectively. Therefore:

$$\sigma_P = \sqrt{2}\,\sigma_n \cdot \sqrt{2A^2 + \sigma_n^2} , \qquad (8.19)$$

The relationship (8.19) shows that the the error increases also with the signal amplitude and not only with the noise power. Nevertheless, the *relative* error tends to zero for increasing signal amplitude, as it was to expect:

$$\lim_{A \to \infty} \frac{\sigma_P}{\bar{P}} = \lim_{A \to \infty} \frac{\sqrt{2}\sigma_n \sqrt{2A^2 + \sigma_n^2}}{A^2 + \sigma_n^2} = 0 \qquad (8.20)$$

The relative error on the estimated power is:

$$\varepsilon_P = \frac{\sigma_P}{\bar{P}} = \frac{\sqrt{2}\sigma_n \sqrt{2A^2 + \sigma_n^2}}{A^2 + \sigma_n^2} \simeq \frac{2\sigma_n}{A} = \frac{\sqrt{2}}{\sqrt{SNR}} \quad (A \gg \sigma_n) \qquad (8.21)$$

The hypothesis of high SNR ($A \gg \sigma_n$) allowed us to assume Gaussian distribution for the amplitude and phase of the received signal, which, in turn, makes it possible to express the statistics of these quantities in analytic form. This assumption should not be regarded as limiting, since, even at low sounding heights, the iterative approach of the OL SW (i.e., the progressive reduction of the bandwidth) still guarantees "good" values of the *signal-to-noise* ratio (typically, $SNR > 10\,dB$ for sounding heights in the range of 38 - 36 km, (super-refraction height $\simeq 33\,km$)). Furthermore, for *signal-to-noise* ratios worse than $10\,dB$, the error analysis does not provide much information. This remark is based on the observation that the calculated uncertainties of the signal parameters are functions of the SNR (see equations (8.12), (8.13), (8.15), and (8.21)). Therefore, since the SNR is not a deterministic quantity, but rather a random variable, whose current value must be estimated from the data, the error committed on the retrieval of the signal parameters can be specified only within a margin of uncertainty (which can, therefore, be regarded as the uncertainty of the uncertainty!).

As an example[6], a theoretic value of the SNR of $10\,dB$ leads to a theoretic uncertainty on the power data $\varepsilon_P \simeq 45\%$ (see eq. (8.21)). This implies that, when computing the SNR from the observed signal power, the obtained value will be itself affected by error. In the case of the example considered, assuming the value of the observation as the center of the $\pm\sigma$ uncertainty interval, from the power uncertainty $\varepsilon_P \simeq 45\%$ we obtain $SNR^{min} \simeq 7.4\,dB$ and $SNR^{MAX} = 12\,dB$, which leads, respectively, to $\varepsilon_P^{MAX} \simeq 52\%$ and $\varepsilon_P^{min} \simeq 41\%$, that constitute the limit values of the uncertainty interval of the specified power error. By making use of eq. (8.22), it is possible to obtain the correspondent uncertainty of the amplitude, which is $\varepsilon_A \simeq 22.5\% \pm 2.5\%$, i.e., the amplitude uncertainty is comprised between 20% and 25%. Clearly, lower values of the SNR would lead to even larger values of the uncertainty interval, which makes the error estimate less meaningful.

This concept shall be better clarified in the next paragraph, where numeric examples of the inaccuracy obtained for different values of the SNR are shown in tab. 8.1.

Estimation of the underlying statistics from the data

Hereafter we present the computation method which was adopted for the estimate of the uncertainty affecting the retrieved frequency- and power profiles.

The statistics discussed in the previous paragraph are theoretical models which are valid for an *infinite* number of samples. When inferring statistics from the data, because of the *finite* number of samples at disposal, the obtained values differ from

---

[6]In this example, the uncertainty on the estimate of the noise level is retained negligible.

the theoretic ones by quantities which become increasingly smaller for increasing number of observations.[7]

As a consequence, the theoretic error obtained by eqs. (8.12), (8.15), and (8.21) does not coincide with the actual estimation error, since the value of the SNR is not exact, but, in turn, must be estimated from the data. Therefore, the error committed in the computation of frequency-, amplitude-, and power values is specified as an estimated quantity, respectively, $\hat{\sigma}_f$, $\hat{\varepsilon}_A$, and $\hat{\varepsilon}_P$. Each of these quantities depends on the estimated *signal-to-noise* ratio, $\widehat{SNR}$:

$$\hat{\varepsilon}_A = \frac{\sigma_n}{\hat{A}} = \frac{1}{\sqrt{2}} \cdot \frac{1}{\sqrt{\widehat{SNR}}}, \tag{8.22}$$

$$\hat{\sigma}_f = \frac{1}{2\pi \Delta t} \cdot \frac{1}{\sqrt{2}} \cdot \frac{1}{\sqrt{\widehat{SNR}}}, \tag{8.23}$$

$$\hat{\varepsilon}_P \simeq \frac{2\sigma_n}{\hat{A}} = \frac{\sqrt{2}}{\sqrt{\widehat{SNR}}} \qquad (A >> \sigma_n) \tag{8.24}$$

It follows that the starting point of the error analysis is the estimation of the *signal-to-noise* ratio, which shall provide the quantity $\widehat{SNR}$.

To this purpose, first of all the base-line noise is evaluated by following the procedure described in par. 8.1.1. As already mentioned, since the analysis is conducted for high values of the SNR, the fluctuations of the noise power density are small in comparison with the signal power, so that the noise power density is assumed constant over time. Furthermore, in order to let the uncertainty in the estimate of the SNR depend only on the estimate of the signal, the noise statistics, once estimated, are assumed as *known* (i.e., $\hat{\sigma}_n^2 = \sigma_n^2$).

The second quantity needed for the computation of $\widehat{SNR}$ is the (estimated)

---
[7]The theoretic distribution represents the *limit* which the inferred (estimated) statistics tend to.

signal amplitude, which can be obtained by means of the *sample mean*:

$$\hat{A} = \frac{A_1 + A_2 + \cdots + A_n}{n} , \tag{8.25}$$

where the values A_i of the samples are realizations of random variables.

Also the variance of the underlying distribution is, strictly speaking, not known and must be estimated from the data. A common estimator of the variance of the data is the (adjusted) *sample variance*:

$$\hat{\sigma}_{\hat{A}}^2 = \frac{(A_1 - A)^2 + (A_2 - \hat{A})^2 + \cdots + (A_n - \hat{A})^2}{n-1} \tag{8.26}$$

Clearly, $\hat{A}$ and $\hat{\sigma}_{\hat{A}}^2$ are random variables. If the r.v. A_i are uncorrelated with $\mathscr{E}\{A_i\} = A$, and $\sigma_{A_i}^2 = \sigma_n^2$, it can be shown that ([74]):

$$\mathscr{E}\{\hat{A}\} = A , \tag{8.27}$$

and

$$\mathscr{E}\{\hat{\sigma}_{\hat{A}}^2\} = \sigma_n^2 , \tag{8.28}$$

that is, the sample mean and the adjusted sample standard deviation are *unbiased* estimators.

Fig. 8.4 shows the estimate of the *Signal-to-Noise density* ratio $\widehat{SNR_0}$,[8] calculated by subtracting (logarithmic units) the estimated noise power density (baseline noise power density, $\hat{N}_{0|dB} = -182.17\,dBm/Hz$) from the calculated signal intensity (see fig. 7.13)[9] [10]. In order to obtain the SNR, the system bandwidth B must be

---
[8] Ratio of signal power to noise spectral power density, or *Signal-to-Noise* ratio referred to unity bandwidth.

[9] The observed values of the signal amplitude were not averaged accordingly eq. (8.25), since the action of the implemented low-pass filter was retained satisfactory (cut-off frequency: $0.5\,Hz$).

[10] The estimated noise was subtracted (linear units) from the observed values of the signal intensity prior to perform the calculation of the SNR. Even if the effect is negligible for $SNR > 0\,dB$,

considered:

$$\widehat{SNR} = \frac{\hat{P}}{2\sigma_n^2} = \frac{\hat{P}}{2\hat{N}_0 B} = \widehat{SNR}_0 \cdot \frac{1}{B} \ , \tag{8.29}$$

where N_0 is the estimated noise spectral power density. For the specific case considered, $B = 0.5\,Hz$, so that $SNR = 2 \cdot \widehat{SNR}_0$.

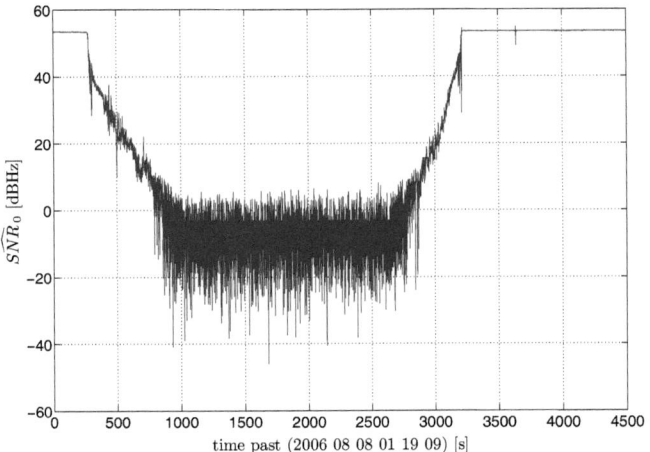

Figure 8.4: *Estimate of the signal-to-noise density ratio (X-Band) for the VEX-VeRa occultation pass recored on DoY220-2006, orbit #109 at the NNO G/S.*

Once $\widehat{SNR}$ has been computed from the obtained estimate of noise power density and signal amplitude, the uncertainty on the other signal parameters can be computed[11] (see eqs.(8.22), (8.23), and (8.24)):

$$\hat{\varepsilon}_P \simeq \frac{2\sigma_n}{\hat{A}} = \frac{\sqrt{2}}{\sqrt{\widehat{SNR}}} = 2\hat{\varepsilon}_A \ , \tag{8.30}$$

---

this operation is conceptually necessary, since $SNR = \frac{S}{N} \neq \frac{S+N}{N}$.

[11]The standard deviation of the noise was assumed *tout court* as the uncertainty for the amplitude (justified by eq. (8.28)), which allowed to avoid the computation of the sample variance.

$$\hat{\sigma}_f = \frac{1}{2\pi\Delta t} \cdot \frac{1}{\sqrt{2}} \cdot \frac{1}{\sqrt{\widehat{SNR}}} = \frac{\hat{\varepsilon}_A}{2\pi\Delta t} \tag{8.31}$$

As an example, table 8.1 presents the inaccuracy of the computed profiles of amplitude, power, and frequency for the VEX-VeRa occultation pass on DoY220-2006, orbit #109. The table shows the error calculated for three different atmospheric penetration depths of the X-Band carrier signal in the Venus atmosphere. The sampling time was $\Delta t = 0.1\,s$.

$(A \gg \sigma_n)$ $(\hat{\sigma}_{\hat{A}} = \sigma_n)$	height = 120 km $\widehat{SNR} = 57\,dB$	height = 46 km $\widehat{SNR} = 30\,dB$	height = 38 km $\widehat{SNR} = 10\,dB$
$\hat{\varepsilon}_A = \frac{\hat{\sigma}_{\hat{A}}}{\hat{A}} = 1/\sqrt{2\widehat{SNR}}$	0.1%	2.2%	22.4%
$\hat{\varepsilon}_P = \frac{\hat{\sigma}_P}{\hat{P}} = \sqrt{2}/\sqrt{\widehat{SNR}}$	0.2%	4.5%	44.7%
$\hat{\sigma}_f = \frac{1}{2\pi\Delta t} \cdot \frac{1}{\sqrt{2}} \cdot \frac{1}{\sqrt{\widehat{SNR}}}$	$1.6\,mHz$	$35.6\,mHz$	$356\,mHz$

Table 8.1: *Inaccuracy of the calculated amplitude, power, and frequency, specified for three different atmospheric penetration depths of the S/C carrier signal recorded during the VEX-VeRa occultation pass on DoY220-2006, orbit #109 at the NNO G/S. (X-Band).*

Critical interpretation of the performed error analysis

The frequency- and power profiles retrieved from OL occultation data constitute the so-called "level 2" data (L2)[12]. As illustrated in Chapter 5, they represent an intermediate processing level needed for the retrieval of physical quantities, such as temperature and pressure.

As a consequence, conclusions can be drawn from the performed error analysis only by addressing the relationships among the quantities intervening in the different levels of the processing chain and observing how the error propagates. As a matter of fact, this can only be done at the end of the whole data evaluation process.

---

[12] Raw data, i.e., voltage samples, are indicated as "level 1" (L1-data).

Although a thoroughly analysis of the error propagation through the evaluation process is beyond the scopes of this work, in the following we would like to briefly illustrate the guidelines of such analysis and, by referring to the work of Lipa and Tyler ([64]), to give an evaluation of the effects of the uncertainties computed in par. 8.1.2 on temperature- and pressure profiles, which will be derived from OL L2-data.

We start by considering that the inaccuracy of the retrieved *frequency* profile reflects itself in the inaccuracy of the determination of the spacecraft *velocity* components in the direction of the asymptotes, which, in turn, leads to an error in the computed *bending angle* (see eq. (5.1)), and, therefore, in the computation of the *impact parameter* (see eq. (5.2)). The error on the retrieved profile of the bending angle vs. impact parameter enters the Abel inversion (see eq. (5.4)) determining a correspondent error on the retrieved *refractive index* (or, equivalently, on the refractivity). Through eq. (5.16), the error propagates to the retrieved temperature- (eq. (5.18) and pressure profiles (equations (5.17) and (5.19))).

In the frame of the NASA mission *Mariner 10*, an analysis of statistical and computational uncertainties affecting the retrieved Venus atmospheric profiles was performed by Lipa and Tyler ([64]). Applying a different method for the reduction of OL data from radio occultation measurements, they indicated as typical values of the precision of frequency measurements $10^{-3}\,Hz$ prior to occultation entry, and $10^{-2}\,Hz$ well into the atmosphere. These values are in agreement with those reported in tab. 8.1.

From the statistical uncertainty of the retrieved frequency profiles due to receiver thermal noise, Lipa and Tyler ([64]) computed standard deviations of temperature and pressure to be respectively $< 2\,K$ and $2\,mbar$, over a range of sounded heights 90 - to 37 km.

8.1.3 USO Phase Noise

The phase noise of the reference on-board oscillator, the USO, is a random error which is superimposed on the signal in addition to the thermal noise at the G/S. Phase noise and receiver thermal noise are considered uncorrelated, therefore for the total frequency noise power hold the relationship:

$$\sigma_{tot} = \sqrt{\sigma_f^2 + \sigma_{AV}^2} \quad , \tag{8.32}$$

where σ_{AV} is the frequency error due to the phase noise.

The frequency error σ_{AV} is related to the Allan Deviation (AD) σ_y through the received carrier frequency f_c ([45]):

$$\sigma_{AV} = f_c \cdot \sigma_y \tag{8.33}$$

For the Allan Deviation is assumed:

$$\sigma_y = \frac{\Delta f}{f_c} = 4 \times 10^{-13} \quad , \tag{8.34}$$

which was obtained for integration times intervals comprised between $10\,s$ and $100\,s$ (see App.D.1).

Taking as f the carrier frequency $f_X = 8.4\,GHz$ and $f_S = 2.2\,GHz$, respectively at X- and S-Band, leads to values $\sigma_{AV_X} = 3.36\,mHz$ and $\sigma_{AV_S} = 0.88\,mHz$, respectively at X- and S-Band. Therefore, from tab.8.1 it is possible to see that the effect of the phase noise at X-Band are significant only for very large values of the SNR (free-space propagation), otherwise the contribution of the thermal noise in the relationships (8.32) is predominant. At S-Band, the larger contribution of the thermal noise, due to lower values of the SNR ($\Delta SNR_{X-S} \simeq 23\,dB$ for free-space propagation), and the lower value of the frequency error σ_{AV} (due to the lower frequency of the carrier signal) make the contribution of the phase noise not significant in comparison to thermal noise.

However, a more accurate analysis should point out that the time series of the frequency values obtained in the last processing stage of the OL SW generally retain a sample-rate $\geq 10/s$, with system bandwidth of order of magnitude $1\,Hz$ or smaller. Since the values of these parameters differ from the correspondent values of the Allan Deviation's measurement parameters, assuming for the AD the value specified by eq. (8.34) seems to be to some extent arbitrary.

An alternative way to evaluate the error induced by the oscillator phase noise is offered by the measurement of the phase noise spectral density, considered in the frequency interval $0\,Hz$ - $1\,Hz$. Unfortunately, this is most commonly the frequency interval which is not covered by the measurement, due to the long integration time which would be required at low frequencies. (The integration time goes to infinite when the frequency goes to zero). However, an approximate evaluation could be afforded by extrapolation. Taking as a reference the X-Band measurement reported in fig. D.3, assuming a slope of $10\,dB$ pro decade in the frequency interval $0\,Hz$ - $1\,Hz$, we estimate a *signal-to-phase noise* ratio in the range 57 - $59\,dB$, which is comparable with the X-Band thermal *signal-to-noise* ratio (free-space propagation). Therefore, eq. (8.32) becomes:

$$\sigma_{tot} = \sqrt{2\,\sigma_f^2} \qquad (8.35)$$

Jitter noise

Clocking jitter of the ground-station H_2-Maser can also be interpreted as a phase noise, as it displaces the samples of a random quantity, which leads to an error in the recovered phase. At the adopted sampling frequency $f_s = 10^5$, the frequency stability of the H_2-Maser of the New Norcia G/S, $\Delta f/f = 10^{-15}$, leads to a maximum frequency error $\varepsilon_f = 10^{-10}\,Hz$, which can well be neglected in comparison with thermal noise.

8.1.4 Quantization noise

After down-conversion to IF, the received analog carrier signal undergoes *Analog-to-Digital* conversion (ADC) which consists in sampling the signal at rate of 10^5 sample/s and digitizing the obtained samples at 16-bit quantization resolution ([55]). Due to further operations on the 16-bit word, the actual resolution is 15 bits ([55]). This leads to a width of the quantization interval $q = \Delta/2^{15}$, where Δ is the amplitude excursion of the wave to be digitized (absolute value of the difference between maximum- and minimum value of the amplitude). In the case of harmonic signals is $\Delta = 2$, thus $q = 1/2^{14}$. The variance of to a quantization error is (see App. H.3):

$$\sigma_q^2 = \frac{q^2}{12} \simeq 3 \times 10^{-10} \equiv -95\, dB; \tag{8.36}$$

Since the signal was taken of unity amplitude, this value corresponds to the *Signal-to-Noise* ratio for quantization noise, SNR_q. In order to compare the quantization noise with the thermal noise, the maximum SNR is evaluated. From fig. 8.4, $SNR = SNR_0 \cdot B \equiv 57\, dB$, which implies that the quantization noise is $38\, dB$ below the thermal noise floor, that is, $\sigma_q^2 < 10^{-3} \times \sigma_n^2$. This shows that the quantization noise brings a negligible contribution to the error analysis presented in tab. 8.1.

8.1.5 Fluctuation in the propagation path

Since the occultation experiment is based on an integral measurement which assumes spheric symmetry (the refractivity depends only on the radial distance form the center of the planet (see Ch.5)), possible variations of the refraction index along each "single path" (path in the Venus atmosphere of each ray received at the G/S) can not be resolved.

8.2 Systematic Errors

Systematic errors differ from statistical errors as they have a deterministic origin[13], that is, they are caused by a phenomenon which is reproducible, such as a computational error, a clock drift, a bias in the response curve of a transducer, etc. Systematic errors can be corrected, as long as the origin of the error is known and modeled. Generally, systematic errors reflect themselves in a bias, which becomes evident after reduction of the data. An error in the antenna pointing would show up as supplementary attenuation in the received power profile (see par. 8.2.2), whereas inaccuracy in the reconstruction of the orbit would lead to a non-zero Doppler frequency of the reduced data in segments where no medium effects are present, as the time preceding occultation entry, or following occultation exit (see par. 8.2.4). In

---
[13]Systematic errors can be also regarded as "large-scale uncertainties that are highly correlated over the course of an observation" ([64]).

particular, frequency residual are corrected by subtracting a bias curve fitted on a subset of pre- or post-occultation data points (base-line fit) ([92]).

8.2.1 Inaccuracy in the evaluation of the effects of the Earth ionosphere and troposphere

The neutral atmosphere is a non-dispersive medium with respect to radio waves up to frequencies of $15\,GHz$. The refraction index n, however, is dependent on air pressure, temperature, humidity, and zenith angle. The effect on the signal reaching the G/S is a path delay that reaches 2.0 - 2.5 m in the zenith direction and increases approximately with the cosecant of the elevation angle, yielding about a 20 - 28 m delay at a 5° elevation angle, independently on the frequency. The tropospheric contribution varies with time and will lead therefore also to a measurable additional Doppler effect of the order of $1.5\,mm/s$ (corresponding to an absolute frequency shift of $\sim 40\,mHz$ at X-Band and $\sim 10\,mHz$ at S-Band) during a full day ground station tracking ([45], [40]). The ionospheric refraction leads to a reduction on group velocity and an increase in phase velocity. The refraction index depends on local time and geomagnetic activity. The delay reaches typically 0.1 - 2 ns (3 cm-60 cm) in the zenith direction at $8.4\,GHz$ ([45], [40]).

Further details about the modeling techniques are discussed in [40].

The calibration of the effects induced by Earth ionosphere and troposphere is not among the tasks of the *Open-Loop* SW. This procedure is performed in the next processing level (L3).

8.2.2 Mis-pointing of the S/C antenna

Displacement of the S/C antenna bore-sight from the computed direction (virtual image of the Earth in the Venus atmosphere) affects the received signal strength, since radiated RF power is tapered by the antenna pattern. This is equivalent to an additional attenuation which deteriorates the overall SNR. Measurements performed at X-Band are more sensitive to antenna mis-pointing than S-Band experiments, because the width of the antenna main-lobe diminishes proportionally to the wavelength.

Mostly, mis-pointing is due to small oscillations of the antenna around the nominal pointing. Therefore, the attenuation assumes the form of a ripple modulating the signal amplitude. However, no appreciable effects have been detected so far during data analysis of the VEX mission.

8.2.3 Oscillator drift (aging)

The inherent design of the resonator guarantees an aging rate of 1×10^{-11}/day after a continuous operation of 30 days ([76], see also App.D.1).

A calibration curve for the effects of aging is made available by interpolating the measurements which have been performed throughout the mission.

8.2.4 Inaccuracy in the reconstructed orbit data and in the S/C attitude

As already mentioned, inaccuracy in the reconstruction of the orbit leads to a non-zero Doppler frequency of the reduced data in segments where no medium effects are present. Such a behaviour has been observed in the data. In particular, in proximity of perigee, quadratic terms show up in the frequency drift. The linear trend of the frequency residuals is corrected by subtracting a bias curve fitted on a subset of pre- or post-occultation data points (base-line fit), but it is more difficult to remove quadratic behaviour.

The maximal deviation of the observed quadratic trend from the zero-frequency line is in the order of magnitude of $10^{-1}\,Hz$, reached in portions of the measurements well outside the atmosphere (~ 20 min. after occultation egress, for the observations evaluated so far).

8.2.5 Execution time offset

A time error in the execution of S/C attitude manoeuvres (antenna pointing) impacts the observation in a similar fashion as the mis-pointing error (see above). In fact, pointing the antenna in the correct (computed) direction in advance (or with delay) is equivalent to radiate RF power from a position of the antenna pattern which is displaced from the bore-sight. The sensitivity of the measurements to this error increases with the rapidity of the required antenna slews.

Bistatic-Radar observations are more robust to this kind of error than occultation experiments, due to the large area illuminated by the antenna on ground (antenna foot-print).

Referring to the occultation experiment, in the ingress- and egress phases the height of the antenna beam in the atmosphere changes very fast and, therefore, the antenna slew must be fast in order to compensate for the rapid changes of the refractive index. It implies that these phases of the measurement are more sensitive than the deep occultation phase, where the height of the antenna beam remains almost constant. Furthermore, also the position of the S/C along the orbit determines different sensitivities to a possible error in the manoeuvres execution time: observations carried out at pericenter requires larger S/C slew-rates than measurements performed from the ascending or descending branch.

For the VEX mission, the declared uncertainty in the execution time offset amounts to $\pm 0.5\,s$. Considering a scalar velocity of the S/C $v = 8\,km/s$, this implies an orbital offset of $\pm 4\,km$. Considering again the occultation experiment, the impact of this error strongly depends on the portion of atmosphere being sounded and on the location of the S/C along the orbit (which determines S/C velocity and antenna beam projection in the atmosphere), so that it is not trivial to evaluate the effects of the relevant mis-pointing.

8.2.6 Finite precision representation

Double precision floating-point format (IEEE 754) reserves 11 bits to the exponent and 53 bits to the fraction, which leads to a precision of $log(2^{53}) \simeq 15.955$, that is, 15 decimal digits. This resolution is well below the quantization resolution and therefore is negligible as noise contribution. The achievable precision would only be of concern for the time representation: as the JD at 1th-January-2000 was $2,451,545$ days, the precision is limited to ms ([86]), which would lead to a precision in the determination of the position of the celestial bodies (included the S/C) of $\simeq 10\,m$, that is not satisfactory. The problem could be overcome by introducing the *Modified Julian Date* (MJD), (see par.3.3).

Finite precision in representation of the arrival time of the samples leads to a quantization error of the time stamps of the samples, whose typical *saw-tooth* trend is limited to a maximum error of $60\,ps$, well above the considered accuracy of absolute date (see [54]).

8.2.7 Computational Errors

The time profile of the frequency of the received signal is calculated in the time domain as time derivative of the phase (see par. 7.1.3). Since this operation is implemented numerically, an error arises from the approximation $d\phi/dt \simeq \Delta\phi/\Delta t$. Analogously to analysis of [64] (carried out for the approximation of the derivative of the *bending angle* α w.r.t. the *impact parameter* a), the uncertainty amounts to:

$$\varepsilon = \frac{1}{2}\Delta(d\phi/dt) \tag{8.37}$$

8.2.8 Algorithm

The applied IIR filters present a linear attenuation in the pass-band whose maximum value is reached at the edge of the pass-band (see par. 6.2.6). The maximum value of the attenuation is less than 0.1, which is comparable with the accuracy of the power calibration curve (see par.8.1.1)

Chapter 9

Summary

The motivation for this PhD thesis was the analysis and evaluation of measurement data collected by means of the Radio-Science investigations in the framework of the ESA missions "Venus Express" and "Rosetta". The goal was the conception and development of a software package for the processing of "Open-Loop" (OL) data from radio occultation experiments, and of computational routines for the evaluation of the expected Doppler frequency shift from the surface of planetary bodies and comets arising during "Bistatic-Radar" observations.

After a brief introduction of the ESA missions "Venus Express" (VEX) and "Rosetta" (ROS), the physical principles of Radio Science were discussed along with the different radio science experiments: besides a short insight on the "Solar Corona"-, and "Gravity", experiments, the "Bistatic-Radar" and the "Occultation" sounding methods were described.

The two different techniques, the "Closed Loop" (CL) mode and the Open Loop" (OL) mode, used for signal acquisition and recording on ground in the frame of the VEX radio science occultation experiment, were presented, and the advantages of OL reception mode over the CL reception mode were discussed. It was pointed out that, due to the inherent characteristic of the reception strategy itself (the bandwidth of the tracking circuitry (PLL) can not cope at the same time with high signal dynamics and low values of SNR), CL data have an intrinsic limitation in the maximum depth of sounded height and in the possibility to resolve different signals entering the receiver simultaneously, as in the case of multipath propagation. Furthermore, the vertical resolution achievable by means of CL data is limited to the first Fresnel zone, whereas it has been proven that, if opportunely processed, OL data can reach to sub-Fresnel scale.

The strategy conceived for the reduction of occultation open-loop data and the developed Open-Loop Data Processing Software (OL SW) were discussed in details.

Essential skill of the OL SW is the progressive reduction of the signal bandwidth while at the same time maintaining high time resolution of the data. This implies high spacial resolution of the sounded media (i.e., the Venus atmosphere) and the capability of resolving effects of multipath propagation. Furthermore, if complemented with opportune routines, based on the theories which were presented (i.e., Radio-Holography, and Diffraction Theory), the high information-rate provided by the OL SW can be exploited to resolve structures smaller than the Fresnel scale.

The proposed method is based on an iterative, hybrid approach, which makes use of classical Fourier spectral analysis and time-domain evaluation. In a pre-processing stage, after binary-to-decimal conversion, and correction of the residual vacuum Doppler frequency shift (straight-line), the raw-data (10^5 complex voltage sample per second) are numerically down-converted by a generated mix-signal (based on a model-atmosphere) which mimics the expected frequency shift induced by the atmosphere. The purpose of this operation is to reduce the signal dynamics, that is, the signal bandwidth, thus allowing low-pass filtering and sample decimation with reduction factors of the order of magnitude 10^2, a remarkable advantage in terms of computation velocity and memory allocation. After completion of pre-processing, the iterative processing starts. At each iteration step, an estimation of the signal frequency, based on Fourier analysis (FFT), takes place. The time series provided by the estimation (the "frequency residual") is used to compute the phase of a complex exponential function of unit amplitude which, in turns, is used for numerical down-conversion of the signal itself.

In the iterative execution of this process lies the core concept of the developed program. In fact, mixing the signal by the estimate of its own frequency progressively reduces signal dynamics, which, in turn, allow increased observation windows in the

FFT analysis. This results on one side in a progressive enhancement of frequency resolution, and, on the other side, in the reduction of the noise power in each FFT resolution cell, thus implying a progressive improvement of the SNR. Both effects lead to increasing accuracy in the evaluation of the signal frequency residual at each processing step, so that the *cut-off* frequency of the low-pass filter applied at the end of each step can be progressively reduced. The iteration process ends when the maximum achievable accuracy, set by receiver thermal noise and frequency stability of the on-board ultra-stable oscillator (USO), is reached. Typically, the analysis is accomplished within three to four iteration step. The last down-converted signal is evaluated in the time domain as time derivative of the signal phase in order to obtain *instantaneous* values of the frequency. This requirement is not attainable by means of the FFT, as the calculated spectra result from averaging over the whole observation period T. The mix-signals (frequency estimates) obtained in each iteration step are stored for reconstruction of the total frequency shift at the end of the processing. For this reason, they have to be regarded merely as "arbitrary" shifts which are "undone" at the end of the processing by adding them to the time domain estimate. This is a central feature of the outlined process, since it causes the achievable accuracy of the frequency estimation to depend on the accuracy of the time domain estimate (given by the bandwidth of the last processing step) and not on the various estimates obtained by means of the FFT during the iterative procedure (under assumption that the inaccuracy of the frequency estimate of each step is considered in the evaluation of the cut-off frequency of the low-pass filter).

By means of the developed software package, atmospheric sounding could be extended towards low altitudes, in some cases to the limit of the super-refraction height ($\sim 33\,km$). Also the time resolution was significantly increased w.r.t. CL data (in principle the time resolution of the OL data is only limited by the sampling rate).

It was shown that small-scale structures (i.e. undulations, or sinusoidal features) of the refractive index, associated with temperature inversion layers in the cloud region (70 - 50 km) cause multipath propagation at ingress- and egress phase. As a result, multiple signals, corresponding to "rays" from different heights, enter the ground receiver simultaneously, so that a bending angle can not be univocally attributed to the received frequency. By means of the OL SW, multiple signals were resolved and a procedure has been successfully tested, which re-establishes the correct sequence of the values of the refractive index with monotonically decreasing height.

In order to validate the obtained results, error analysis was performed and the most significant error sources were considered. Analysis of statistical error affecting the processed data shows typical values of the frequency uncertainty of $10^{-3}\,Hz$ prior to occultation entry (free-space propagation), and $10^{-2}\,Hz$ well into the atmosphere. The correspondent uncertainties of the power profiles are $\varepsilon = 0.2\%$ and $\varepsilon = 4.5\%$.

The developed software package was especially designed in order to be applied to the routine data analysis of the VEX-VeRa Occultation data from the first nine occultation seasons and of the same kind of data which are to be obtained from three future occultation seasons, with observations planned through the period 2011 - 2012. Furthermore, the OL SW will be used for processing of the OL data which will be provided by the Radio Science Investigation on-board the Rosetta S/C after beginning of the science observations (early 2014).

First results from the processed OL data have shown some remarkable characteristics; important scientific results are expected from thorough analyses. Further development, optimization, and application of the OL SW package will allow to investigate into more details small-scale structures of the refractive index and fine structures in the temperature inversion regions and multipath regions in the Venus atmosphere (cloud layer) and the investigation of atmospheric waves.

Besides the analysis of occultation OL data, software routines were developed for the calculation of the expected Doppler frequency shift from the surface of planetary bodies and comets arising during "Bistatic-Radar" observations. Comparison of simulation results with measurements performed during the VEX-VeRa bistatic radar observation at Maxwell Montes on June, 15th 2006 shows good agreement, which, therefore, validates the software.

Appendix A

Analog PM and FM modulations[1]

The case of single sinusoidal tone modulating a RF carrier signal will be analyzed in this appendix. In particular, the case of frequency modulation (FM) will be shown in details[2]. The base-band signal to be transmitted is assumed to be of the form:

$$m(t) = \Delta\omega \cdot cos(\omega_m t + \psi) \qquad (A.1)$$

where the constants $\Delta\omega$, ψ, and ω_m represent the amplitude, the phase offset at the initial time ($t=0$) and the angular frequency, respectively. In the FM case, this signal is supposed to directly modulate the carrier frequency ω_c as an additive contribution, so that the instantaneous angular frequency becomes:

$$\omega(t) = \omega_c + m(t) \qquad (A.2)$$

The instantaneous phase is calculated as the time integral of the instantaneous frequency (by neglecting the constant phase offset ψ):

$$\phi(t) = \int_0^t \omega(\tau)d\tau = \int_0^t [\omega_c + m(\tau)]d\tau = \omega_c t + \frac{\Delta\omega}{\omega_m}sin(\omega_m t) \qquad (A.3)$$

The quantity $\Delta\omega/\omega_m$ is given the name of *phase modulation index*. By indicating

---
[1] An extensive treatment of this subject can be found in [17].
[2] Apart from the symbols, the case of phase modulation (PM) is essentially the same.

this quantity as β, the RF modulated signal can be written as:

$$s(t) = \cos\left[\omega_c t + \beta \cdot \sin(\omega_m t)\right] \tag{A.4}$$

Upon making use of the complex notation, the expression of the RF modulated signal becomes:

$$s(t) = \Re\left\{e^{j\omega_c t} \cdot e^{j\beta \sin(\omega_m t)}\right\} \tag{A.5}$$

The base-band complex signal is indicated by underlining the letter which represents the RF modulated signal:

$$\underline{s(t)} = e^{j\beta \sin(\omega_m t)} \tag{A.6}$$

Substituting the instantaneous phase of the base-band signal ($\omega_m t$) by the Greek letter ξ gives a a periodic (complex) function of the variable ξ, $f(\xi) = exp\{j\beta sin(\xi)\}$, whose Fourier series expansion over the interval $[-\pi, \pi]$ is:

$$f(\xi) = \sum_{k=-\infty}^{\infty} c_n e^{-j2\pi \frac{n}{2\pi} \xi} \tag{A.7}$$

The Fourier coefficients must be function of the parameter β:

$$c_n = J_n(\beta) = \frac{1}{2\pi} \int_{-\pi}^{\pi} e^{j\beta \sin(\xi)} e^{-jn\xi} d\xi \tag{A.8}$$

The functions $J_n(\beta)$ are called *Bessel function of the first kind of order n*. It holds for the the base-band complex signal :

$$\underline{s(t)} = \sum_{k=-\infty}^{\infty} J_n(\beta) e^{-jn\xi} \tag{A.9}$$

It can be recognized, by taking the Fourier transform of the above, that the power spectrum of such a signal is made up of infinite Dirac pulses, whose area is given by the relevant Bessel function squared. The Bessel functions have the following symmetric property :

$$J_{-n}(\beta) = (-1)^n \cdot J_n(\beta) \tag{A.10}$$

For this reason the power spectrum is even-symmetric around the origin (or around the carrier frequency when the base-band signal modulates the carrier). The single Dirac pairs are called *side-bands*. In most practical cases, the value of the modulation index is chosen in the proximity of the unity, so that the most relevant contribution to the signal power is given by the first side-band, as it can be seen in fig.A.1. This allows setting a minimum receiver bandwidth on the receiver side, thus minimizing the contribution of the thermal noise.

For the evaluation of the bandwidth occupation BW of the signal A.4, the "Carson" formula holds (baseband):

$$BW = 2(\Delta\omega + \omega_m) \tag{A.11}$$

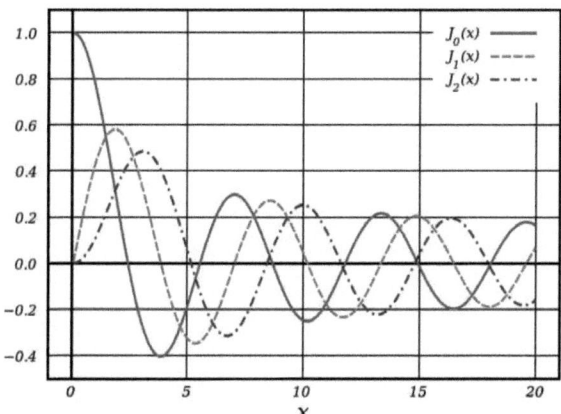

Figure A.1: *Bessel's function of first kind of order 1 to 3.*

Appendix B

The Signal-to-Noise Ratio (SNR)

The power of the received signal at the input stage of the receiver is related to the power of the thermal noise measured at the same point of the receiving chain by the dimensionless quantity:

$$SNR = \frac{P_R}{N} \tag{B.1}$$

where P_R and N denote the power of the received signal and of the thermal noise, respectively. This quantity is given the name of *Signal-to-Noise Ratio (SNR or S/N)* and it is representative of the radio-link quality.

We now intend to analyze each of the two terms on the right side of (B.1), starting by P_R. When a transmission between two antennas occurs, the power of the received signal P_R depends on the relative position of both antennas. Being P_T the power of the transmitted signal and R the distance separating the antennas, the power of the received signal is given by:[1]

$$P_R(\theta_r, \phi_r) = \frac{P_T G_T(\theta_t, \phi_t)}{4\pi R^2} \cdot A_e(\theta_r, \phi_r) \tag{B.2}$$

where the azimuthal angle θ_r and the elevation angle ϕ_r give the direction of the transmitting source in the coordinate system of the receiving antenna and vice versa

---

[1] In (B.1) the dependency of P_R on the direction of arrival (given by the angles θ_r, ϕ_r) was dropped, as it is assumed that it is the direction of maximum power transfer.

for θ_t and ϕ_t.[2]

The function $G_T(\theta, \phi)$ is called *gain*. It is a dimensionless coefficient which describes the *radiation pattern* of the antenna, i.e. the way the antenna distributes the radiated electromagnetic power in space. An antenna which radiates electromagnetic energy uniformly in each space direction is said to be *omni-directional* or *isotropic*.[3] It is possible to intervene on the design of the antenna in order to obtain more radiated power in a certain direction. In this case the emitted power shall be less under other directions, as the radiated power distributed on a spherical surface centered at the transmitting antenna at any distance R[4] must equal the total transmitted power. Typically, for a given value of the variables θ, ϕ, the gain function has an absolute maximum, which is indicated as G. The product of the transmitted power for the maximum gain is given the name *Equivalent Isotropically Radiated Power* ($EIRP$): it represents the power that an isotropic radiator should emit to produce the observed power in the direction of maximum antenna gain. The value of the gain, when expressed in logarithmic units, is indicated in dBi (decibels over isotropic).

Whereas the first term on the right side of (B.2) represents the surface density of the transmitted power at given distance R from the radiating source and has the dimension $\left[\frac{W}{m^2}\right]$, the second term on the right side of (B.2) is called *effective aperture* of the receiving antenna. This parameter has the dimension $[m^2]$ [5], but it not necessarily related with the physical surface of the antenna. In the following it shall be assumed that the receiving antenna is oriented for maximum response (i.e. the plane containing the antenna is perpendicular to the propagation direction of the incoming electromagnetic wave) and the incoming electromagnetic wave is conveniently polarized. Let's first of all define the aperture A as the area given by

$$A = \frac{P_L}{S} \tag{B.3}$$

where P_L is the fraction of the transmitted power "captured" by the receiving

---

[2] The subscripts r, t will be omitted when there is only one coordinate system to be considered.
[3] This an ideal case which is not physically realizable. The best approximation of an ideal isotropic radiator is the Hertzian dipole.
[4] provided $R \gg d$, where d is the extension of the largest antenna dimension (far-field condition).
[5] so that the expression of the power of the received signal $P_R(\theta_r, \phi_r)$ in (B.2) is consistent, having the dimension $[m^2]$

antenna and made available to the load (input impedance of the receiver) and S is the intensity of the Poynting vector of the incident wave (assumed plane) impinging on the receiving antenna, so that A indeed has the dimension of a surface ($[m^2]$)[59]. The power P_L which can be transferred to the load is maximum if the antenna impedance is *matched* to the load (i.e. antenna impedance and load impedance are complex conjugate). If the matching condition holds, the aperture in (B.3) is called *effective* aperture and it is given by[6]

$$A_e = \frac{P_{LM}}{S} \tag{B.4}$$

where P_{LM} is the power transferred to the load when antenna impedance and load are matched. If the antenna has no ohmic losses there is no power which is dissipated as heat in the antenna structure. In this case the effective aperture takes its maximum value A_{em} and it is related to A_e by:

$$A_e = A_{em} \cdot \eta \tag{B.5}$$

where η is called antenna *efficiency* and accounts for ohmic losses in the antenna.

Because of the *Lorentz reciprocity theorem*, which states that antennas show the same characteristics when transmitting and when receiving, more specifically that an antenna's radiation and receiving patterns are identical, it is possible to relate the effective aperture to the maximum antenna gain:

$$A_e = \frac{\lambda^2}{4\pi} \cdot G \tag{B.6}$$

where λ is the wavelength of emitted or received electromagnetic radiation[7].

Now, considering that transmitting and receiving antenna are supposed to be oriented in a way to have the maximum of the radiation pattern pointed in the

---

[6] A_e in (B.4) is the same quantity as $A_e(\theta_r, \phi_r)$ in (B.2) evaluated when the receiving antenna is oriented for maximum response.

[7] The antenna gain $G(\theta, \phi)$ is related to the directivity $D(\theta, \phi)$ by $G(\theta, \phi) = \eta \cdot D(\theta, \phi)$. The directivity is defined as the ratio of the power emitted by the antenna in a certain direction to the the power emitted by an isotropic radiator in the same direction.

direction of the other, the power at the input impedance of the receiver in (B.2) can be written as:

$$P_R = \frac{P_T G_T}{4\pi R^2} \cdot A_e \tag{B.7}$$

Adopting the subscript R for the gain in (B.6) and substituting (B.6) in (B.7) yields:

$$\frac{P_R}{P_T} = \left(\frac{\lambda}{4\pi R}\right)^2 \cdot G_T G_R \tag{B.8}$$

where G_T and G_R are the gain of the transmitting and the receiving antenna, respectively. The (B.8) is known as *Friis's transmission formula*[99]. The reciprocal of the first term on the right side of the (B.8) is commonly related to as *free-space transmission loss* L_{FS}:

$$L_{FS} = \left(\frac{4\pi R}{\lambda}\right)^2 \tag{B.9}$$

In order to explicit the expression of the Signal-to-Noise Ratio in (B.1), we shall now analyze the term N, that is, the power of the thermal noise at the receiver input. It holds:

$$N = K \cdot T_{sys} \cdot B \tag{B.10}$$

where K is the Boltzmann's constant ($1.3806504 \times 10^{-23}$ $\left[\frac{J}{K}\right]$ [70]), T_{sys} is the system temperature in $[K]$, and B is the considered bandwidth in $[Hz]$[8]. The quantity KT_{sys} in (B.10) can be expressed as

$$N_0 = K \cdot T_{sys} \tag{B.11}$$

---

[8]Taking into account that $\left[\frac{J}{K}\right] = \left[\frac{W}{Hz \cdot K}\right]$, multiplying the terms in (B.10) yields the value of the noise power in $[W]$, which is consistent.

N_0 has the dimension $[\frac{W}{Hz}]$: it represent the *noise power spectral density* (one-sided)[9].

The system temperature T_{sys} is a fictive temperature, i.e. the temperature a resistor should have to produce the noise power density N_0 at the receiver ingress. It comprises two different contributions, the *equivalent antenna temperature* T_A and the *input temperature* T_{in}:

$$T_{sys} = T_A + T_{in} \tag{B.12}$$

T_A is the temperature a resistor at the receiver input should have to produce the same noise power density as the one produced by background radiation coming through the antenna. T_{in} represents the noise contribution of the receiver system, referred to the receiver input. The receiver chain is made up of both active and passive elements, which contributes in different ways to the total amount of noise power:

$$T_{in} = T_{in_a} + T_{in_p} \tag{B.13}$$

The noise characteristics of active circuitry, such as amplifiers, are described by the *noise figure F*:

$$T_{in_a} = (F - 1) \cdot T_0 \tag{B.14}$$

where T_0 is the *standard temperature* ($290K$). Passive components, such as cabels, are characterized by the *loss L*:

$$T_{in_p} = (L - 1) \cdot T_0 \tag{B.15}$$

Upon making use of (B.8), (B.9), (B.10), and (B.11), the Signal-to-Noise Ratio

---

[9]The quantity N_0 is also regarded as the power of the noise in a bandwidth of $1Hz$.

in (B.1) can be then expressed as:

$$SNR = \frac{P_T \cdot G_T \cdot G_R}{L_{FS} \cdot N_0 \cdot B} \tag{B.16}$$

As the receiver bandwidth can vary, it is meaningful to refer the link quality to the unit bandwidth, that is, to relate the power of the received signal to the noise spectral power density:

$$SNR_0 = \frac{P_T \cdot G_T \cdot G_R}{L_{FS} \cdot N_0} \tag{B.17}$$

The SNR_0 has the dimension $[Hz]$.

The shown calculation of the SNR refers to the case of pure carrier signal, when no modulating signals are present. When information is transmitted, the total RF power which is available for transmission is distributed between data and carrier signal accordingly to the chosen *modulation index*. For analog angular modulations (FM or PM), the distribution of the total power is ruled by the *Bessel's functions* of the first kind, $J_n(\beta)$, where β is the chosen modulation index (see App. A).

Considering the case of a single sinusoidal tone modulating an RF carrier signal (FM or PM), we shall define the radio-link quality for the carrier signal itself, indicated as S/N, and for the sinusoidal tone, indicated as C/N:

$$\frac{S}{N} = \frac{J_0^2(\beta) \cdot P_T \cdot G_T \cdot G_R}{L_{FS} \cdot N_0 \cdot B_c} \tag{B.18}$$

where B_c is the bandwidth of the receiver set for reception of the carrier signal.[10]

The link-budget calculation for the information signal (i.e. the sinusoidal tone) is:

$$\frac{C}{N} = \frac{2 J_1^2(\beta) \cdot P_T \cdot G_T \cdot G_R}{L_{FS} \cdot N_0 \cdot B_{data}} \tag{B.19}$$

---

[10]Typically the bandwidth of the *phase-lock-loop*, PLL.

where B_{data} is the (two-sided) bandwidth of the receiver, set for proper reception of the data, in this case the transmitted sinusoidal tone[11].

It must be noticed that additional losses, not included in (B.16), (B.17), (B.18), and (B.19) have also to be taken into account when dimensioning a radio link. These degradations can be due to various causes, such as mismatching in the transmission lines (i.e. transmitter-antenna or antenna-receiver), inaccuracy in the antenna pointing or some characteristics of the circuitry, which are not modeled by the system temperature.

---

[11] Choosing a value of β near to one lets the power of the information signal be mostly comprised in the first sideband, so that the receiver bandwidth is minimized. See also App. A

Appendix C

The Allan Variance

The output signal of sinusoidal oscillator disturbed by noise will have the form ([79]):

$$V(t) = [V_0 + \epsilon(t)]\, sin[2\pi f_0 t + \phi(t)] \, , \tag{C.1}$$

where V_0 and f_0 are the nominal voltage and the nominal frequency, respectively, and the terms $\epsilon(t)$ and $\phi(t)$ respectively represent the noise-induced variations of the signal amplitude and phase. Under the assumption of negligible amplitude noise, the "quasi-sinusoidal" oscillator output signal will have the form:

$$V(t) = V_0\, sin[2\pi f_0 t + \phi(t)] \tag{C.2}$$

The instantaneous frequency is defined by:

$$f(t) = \frac{1}{2\pi}\frac{d}{dt}[2\pi f_0 t + \phi(t)] = f_0 + \frac{1}{2\pi}\frac{d\phi(t)}{dt} = f_0 + \Delta f(t) \, , \tag{C.3}$$

where $\Delta f(t)$ is a stochastic process which represents the frequency noise. For stable oscillators must be:

$$|\Delta f(t)| << f_0 \quad \forall t \in \mathbb{R}^+ \tag{C.4}$$

An important parameter, which facilitates the performance comparison of oscillators with different frequencies is represented by the *normalized* frequency variations:

$$y(t) = \frac{\Delta f(t)}{f_0} \tag{C.5}$$

Analogously, it is possible to define a second parameter, which relates the phase variations to the nominal frequency:

$$x(t) = \frac{\phi(t)}{2\pi f_0} \tag{C.6}$$

From eqs. (C.3), (C.5), and (C.6), follows that relationships (C.5) and (C.6) are related through:

$$y(t) = \frac{dx(t)}{dt} \tag{C.7}$$

The Allan Variance (AV) measures the frequency stability of an oscillator. It is defined by making use of the quantity $y(t)$, introduced in the relationship (C.5) ([37]):

$$\sigma_y^2(\tau) = \frac{1}{2(M-1)} \sum_{i=1}^{M-1} [y_{i+1} - y_i]^2 \, , \tag{C.8}$$

where y_i is the ith of M fractional frequency values averaged over the measurement interval τ.

In terms of phase data, the AV may be calculated as:

$$\sigma_y^2(\tau) = \frac{1}{2(N-2)\tau^2} \sum_{i=1}^{N-2} [x_{i+2} - x_{i+1} + x_i]^2 \, , \tag{C.9}$$

where x_i is the ith of $N = M+1$ phase values averaged over the measurement interval τ.

For a general survey of parameters proposed for phase and frequency instability characterization, see [83]

Appendix D

The Ultra Stable Oscillator (USO)

D.1 The USO design

The on-board *Ultra Stable Oscillator* USO (see fig.D.1) provides the *VeRa* radio science experiment on-board Venus Express, and the *Radio Science Instrument*(RSI) on-board Rosetta with the needed frequency stability ($\Delta f/f \sim 10^{-13}$, see after), making available a reference frequency of $\sim 38\,MHz$ for the transponders.

The VEX- and and the Rosetta USOes are based on a unique resonator design "BVA"(*bote vieillage ameliore*). The BVA resonator itself consists of an electrodeless, "SC"(stress compensated) - cut, $5MHz$ 3rd overtone crystal resonator, which is decoupled from its mounting structure by four rigid quartz bridges machined from a monolithic block. This design consists of a sandwich of three crystal plates rigidly clamped together. In contrast to conventional designs, the metalization is deposited on the non-resonating outer two elements, whereas the resonator itself is pure quartz capacitively coupled to the outer electrodes. This design eliminates contamination problems linked to ion migration in the resonator and reduces constraints on the mounting structure.

To achieve the requested frequency, a synthesizer (see Fig. D.2) is needed. There is no numerically controlled oscillator and no software or microprocessor in this design. All division and multiplication ratios are fixed by hard-wired design.

The inherent design of the resonator guarantees an aging rate of 1×10^{-11}/day after a continuous operation of 30 days. The USO is further characterized by an extreme low phase noise, resulting in an excellent frequency stability (Allan Deviation $\sim 4 \times 10^{-13}$ for integration times comprised between $10\,s$ - $100\,s$, see App. C

Figure D.1: *The* VeRa *Ultra Stable Oscillator - USO. (Manufacturer TIMETECH GmbH).*

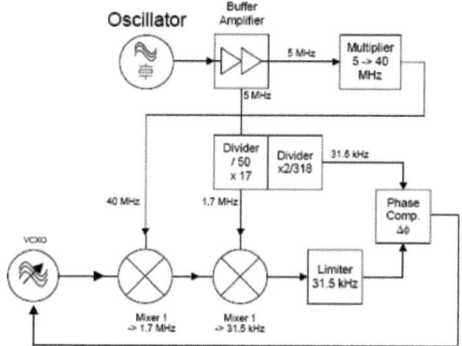

Figure D.2: *Schematic view of the 5 MHz to 38.2 MHz synthesizer of the VEX USO.* ([76])

and D.2). Its dimensions are $16.1\,cm \times 13.0\,cm \times 13.0\,cm$. Its mass is $1.5\,kg$ and it has a power consumption of $\sim 5\,W$. The RF-output of the USO can be muted by a special command while the ultra stable oscillator is still powered ([76]).

D.2 The calibration measurements

The calibration measurements, performed on unit-, subsystem-, and system level, had the purpose to verify that the USO transponder system delivers the RF signal with an adequate purity concerning frequency stability, phase noise, and group delay. The test method relies on the phase detector method and the usage of ultra stable quartz oscillators installed as frequency reference sources in the *VeRa* EGSE (*Electrical Ground Support Equipment*). The test set up and the test results are described in detail in [36], [41], [42], [67].

A fundamental parameter which characterizes the stability of an oscillator is the *phase noise*, whose spectral density, or *phase noise spectrum*, is indicated as $\mathscr{L}(f)$ (one-sided). As an example, fig. D.3 shows a phase noise measurement of the VEX Flight Model (FM) USO driving TRSP#2 in the One Way mode (X-Band). The spectrum shows a clean behavior close to the carrier in the frequency range $f < 50\,Hz$ without contamination by spurious signals (after elimination of $50\,Hz$ harmonics). At higher frequencies, transponder - spacecraft emissions are present, however with a negligible total power contribution. From the presented test it is possible to conclude, for example, that the bistatic radar experiments, which investigate the spectral behavior of the scattered radio signal over a wide frequency range, can be successfully conducted with the USO as reference frequency source for the ONE-D transmission mode.

The parameter which is most closely related to the frequency stability of the RF carrier is the Allan Deviation (AD), or its squared value, the Allan Variance (AV, see App. C). Fig. D.4 shows the result of the Allan Deviation measurement performed on unit level, i.e. on the $38\,MHz$ output signal of the VEX FM-USO. The value of the frequency stability is $\Delta f/f = 5 \times 10^{-13}$ @ $\tau = 100\,s$ (where τ is the integration time). Fig. D.5 shows the result of the Allan Deviation measurement performed on system level, i.e. the X-Band signal at $8.4\,GHz$ generated by the subsystem comprised of the transponder #2 and the USO. The value of the frequency stability is $\Delta f/f = 3 \times 10^{-13}$ @ $\tau = 100\,s$ (where τ is the integration time).

The relationship between the phase noise σ_ϕ^2 and the Allan Variance σ_y^2 can be derived upon integration over a definite frequency interval of the phase noise

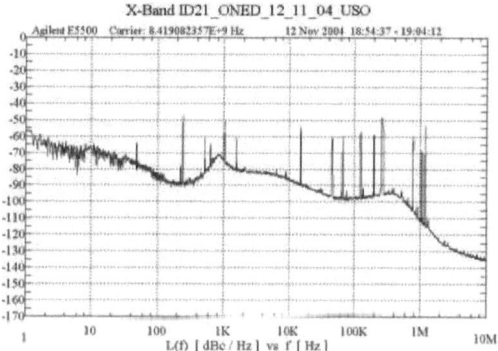

Figure D.3: *Phase noise spectrum of the VEX Flight Model (FM) USO and TRSP#2 in the ONE-D mode at X-Band.* ([67])

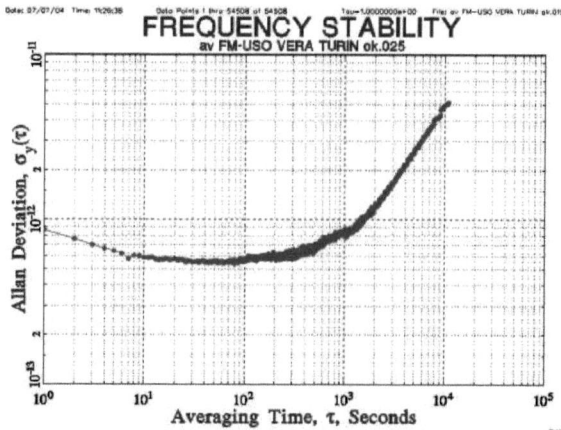

Figure D.4: *Frequency stability measurement of the VEX FM-USO. The graphic represents the Allan Deviation. Frequency values not corrected against drift effects.* ([41])

spectrum (see [79]):

$$\sigma_\phi^2 = \int_0^{f_1} \mathscr{L}(f)\,df = \frac{\sigma_v^2}{\frac{1}{2^2}\left(\frac{\sqrt{2}c}{2\pi f \Delta t}\right)^2} \; , \qquad (D.1)$$

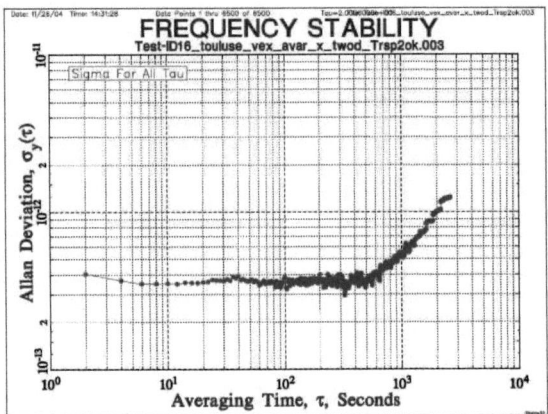

Figure D.5: *Frequency stability measurement of the VEX USO - TRSP#2 subsystem in the ONE-D configuration at X-Band. The graphic represents the Allan Deviation. "Dead-time" corrected data. USO warm-up time approx. 11hrs. Frequency values not corrected against drift effects. ([67])*

where c is the speed of light in vacuum, Δt is the measurement interval, and σ_v is the variance of the velocity error caused by the frequency instability, which can be expressed as function of the Allan Deviation σ_y as:

$$\sigma_v = \frac{c}{\sqrt{2}} \sigma_y = \frac{c}{\sqrt{2}} \frac{\sqrt{2}}{f_0} \sigma_f \ , \qquad (D.2)$$

with σ_f the standard deviation of the measured frequency f_0.

Appendix E

Refraction and reflection of plane waves[1]

In the ideal case of a plane electromagnetic wave imping on a endless and smooth surface separating two physical media with different values of their *refraction index*,[2] part of the energy is reflected back to the medium of provenience and part is transmitted in the other medium.

Referring to the geometry represented in fig. E.1 one defines the incidence angle θ_i as the angle between the wave vector of the incident radiation, $\vec{\kappa}_i$, and the unit vector normal to the surface of incidence, $\hat{n}$. These vectors also define the *incident plane*. Similarly one defines the reflection angle θ_r as the angle between the wave vector of the reflected energy, $\vec{\kappa}_r$, and the unit vector $\hat{n}$. These vectors also define the *observation plane*. In the considered ideal case *specular reflection* takes place at the interface between the two media. Incident plane and the observation plane coincide and for the reflection angle holds:

$$\theta_r = \theta_i \qquad (E.1)$$

Upon crossing the interface between the considered physical media the incident wave changes its propagation direction because of the different propagation velocity

---

[1] An extensive treatment of this subject can be found in [99].
[2] The refraction index is the ratio of the speed of light in vacuum (c) relative to the speed light through the given medium (v). It is: $n = c/v = \frac{1}{\sqrt{\epsilon_0 \mu_0}} \cdot \sqrt{\epsilon \mu} = \sqrt{\epsilon_r \mu_r}$, where the *permittivity* ϵ is the product of the *vacuum permittivity* ϵ_0 and the *relative permittivity* (or *dielectric constant*) ϵ_r. Similarly, the *permeability* μ is the product of the *vacuum permeability* μ_0 and the *relative permeability* μ_r.

in the media. This phenomenon is called *refraction*. It holds (Snell's law):

$$\frac{sin\theta_i}{sin\theta_t} = \frac{n_2}{n_1} \qquad (E.2)$$

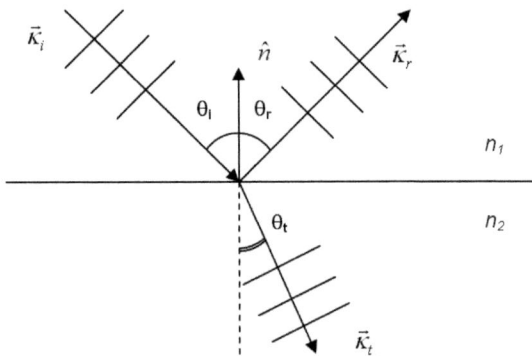

Figure E.1: *Reflected and refracted waves originating from a plane wave impinging on a endless smooth surface separating two media of different refraction index. The incidence angle θ_i is the angle between the wave vector $\vec{\kappa}_i$ of the incident wave and the normal $\hat{n}$ to the interface; the reflection angle θ_r is the angle between the wave vector $\vec{\kappa}_r$ of the reflected wave and $\hat{n}$; the transmission angle θ_t is the angle between the wave vector $\vec{\kappa}_t$ of the transmitted wave and the opposite to the normal $(-\hat{n})$.*

In order to asses the amount of the reflected-, and transmitted radiation, it must be taken into account that the behavior of the electromagnetic fields at the separation surface depends on the incidence angle, on the dielectric and magnetic properties of the considered media, and on the *polarization* of the incident wave[3]. The relationship among these quantities is expressed in the *Fresnel coefficients*,

---

[3]A linearly polarized electromagnetic wave having the $\vec{E}$ field perpendicular to the plan of incidence is said to be *horizontally* polarized (as it is parallel to the surface of incidence), or a **TE** wave. If the $\vec{E}$ field is parallel to the plane of incidence, the wave is said to be *vertically* polarized, or a **TM** wave. A linearly polarized wave with any arbitrary orientation of the $\vec{E}$ field w.r.t. the incidence surface can be decomposed in the sum of a TE and TM wave.

which relate reflected and transmitted fields to the incident radiation:

$$R_\perp = \frac{\eta_2 cos\theta_1 - \eta_1 cos\theta_2}{\eta_2 cos\theta_1 + \eta_1 cos\theta_2} \tag{E.3}$$

$$T_\perp = \frac{2\eta_2 cos\theta_1}{\eta_2 cos\theta_1 + \eta_1 cos\theta_2} \tag{E.4}$$

$$R_\| = \frac{\eta_1 cos\theta_1 - \eta_2 cos\theta_2}{\eta_1 cos\theta_1 + \eta_2 cos\theta_2} \tag{E.5}$$

$$T_\| = \frac{2\eta_1 cos\theta_1}{\eta_1 cos\theta_1 + \eta_2 cos\theta_2} \tag{E.6}$$

where the subscripts "$\|$" and "$\perp$" stand for horizontal-, and vertical polarization, respectively. $\eta_{1,2}$ represents the *characteristic impedance* of the medium and equals the square root of the ratio of the permeability μ to the permittivity ϵ. In case of conducting media, the dielectric constant also has an imaginary part[4], so that the reflection and transmission coefficients are generally complex numbers.

The power reflection coefficient, or *reflectivity* is the squared absolute value of the amplitude reflection coefficient:

$$\Gamma_\perp = |R_\perp|^2 \tag{E.7}$$

whereas the power transmission coefficient, or *transmissivity* is:

$$\Upsilon_\perp = |T_\perp|^2 \tag{E.8}$$

---

[4] In order to account for energy dissipation in conducting media, a complex dielectric constant is introduced, which is $\epsilon_c = \epsilon - j\frac{\sigma}{\omega} = \epsilon' - j\epsilon''$.

For the conservation of the energy must be:

$$\Gamma_\perp + \Upsilon_\perp = 1 \tag{E.9}$$

In the case of vertically polarized incident wave (TM) and two lossless media, there is a value of the incidence angle, called the *Brewster* angle, θ_B, at which total transmission occurs ($R_\parallel = 0$). From (E.6) follows ($R_\parallel = 0$ and $\mu_{r_{1,2}} = 1$):

$$\sqrt{\epsilon_2}\cos\theta_B = \sqrt{\epsilon_1}\cos\theta_2 \tag{E.10}$$

and

$$\tan\theta_B = \sqrt{\frac{\epsilon_2}{\epsilon_1}} \tag{E.11}$$

follows.

Appendix F

Probability, Random Variables, and Stochastic Processes[1]

F.1 Probability

A *non-deterministic* or *stochastic* phenomenon, or process, is one which is not predictable. It depends on random factors, or it is random itself. Such phenomena must be described by means of the *probability theory*.

We define an experiment $\mathfrak{E}$, its outcomes ζ_i ($i \in \mathbb{N}$), and the space $\mathscr{S}$, formed by the outcomes. A set of outcomes $\mathscr{A} = \{\zeta_1, \zeta_2, ..., \zeta_n\}$, forming a subset of $\mathscr{S}$, is called *event*.

The concept of *probability*, following the *axiomatic definition*, is specified as follow: the probability of an event is a number $P(\mathscr{A})$ assigned to this event and obeying the following three postulates:

I. $P(\mathscr{A}) \geq 0$

II. $P(\mathscr{S}) = 1$

III. $P(\mathscr{A} + \mathscr{B}) = P(\mathscr{A}) + P(\mathscr{B})$

(if $\mathscr{A}$ and $\mathscr{B}$ are mutually exclusive; otherwise is: $P(\mathscr{A} + \mathscr{B}) = P(\mathscr{A}) + P(\mathscr{B}) - P(\mathscr{A}\mathscr{B}) \leq P(\mathscr{A}) + P(\mathscr{B})$).

---
[1] An extensive treatment of this subject can be found in [74].

F.2 The random variable

Given an experiment $\mathfrak{E}$ and its outcomes ζ_i, forming the space $\mathscr{S}$, a *random variable* is the relationship which, according to same rules, assigns to every ζ the number $\mathbf{x}(\zeta)$. Given a number x, the notation $\{\mathbf{x} \leq x\}$ represents the set of all outcomes ζ such as $\mathbf{x}(\zeta) \leq x$. This set is an event, and, as such, provided with a probability. Since this probability depends on the number x, it is possible to define a function, the *Distribution Function*, which expresses the probability that the number $\mathbf{x}(\zeta)$ assigned to the event ζ be smaller than, or equal to the number x:

$$D_{(\mathbf{x})}(x) = P\{\mathbf{x} \leq x\} \tag{F.1}$$

The derivative of the distribution function w.r.t. x is called the *Probability Density Function* (pdf):

$$\frac{d\,D_{(\mathbf{x})}(x)}{dx} = f_{(\mathbf{x})}(x) \tag{F.2}$$

F.2.1 Moments of a random variable

In the following, only the case of continuous, real random variables will be considered.

The generic n^{th}-moment of a random variable (r.v.) $\mathbf{x}$ is defined as:

$$m_n = E\{\mathbf{x}^n\} = \int_{-\infty}^{\infty} x^n f(x)\, dx, \tag{F.3}$$

where the "E" operator is called, "expectation"; it has the property of linearity. In particular, the first-order moment is called *expected value*, or *mean* of the r.v.:

$$m_1 = E\{\mathbf{x}\} = \int_{-\infty}^{\infty} x f(x)\, dx = \eta_x \tag{F.4}$$

The generic *central* n^{th}-moment of a random variable (r.v.) **x** is defined as:

$$\mu_n = E\{(\mathbf{x} - \eta_x)^n\} = \int_{-\infty}^{\infty} (x - \eta_x)^n f(x)\, dx \tag{F.5}$$

In particular, the second central moment is called the *variance* of the r.v.:

$$\sigma^2 = E\{(\mathbf{x} - \eta_x)^2\} \tag{F.6}$$

For the linearity of the expectation is:

$$\sigma^2 = E\{(\mathbf{x} - \eta)^2\} = E\{\mathbf{x}^2 - 2\eta_x \mathbf{x} + \eta_x^2\} =$$
$$= E\{\mathbf{x}^2\} - 2\eta_x E\{\mathbf{x}\} + \eta_x^2 = m_2 - m_1^2 \tag{F.7}$$

F.2.2 Joint moments

In the following, the case of two real random variables is considered, as it is propaedeutic to the theory of stochastic processes (see par. F.3).

The generic $(k+r)^{th}$-order joint moment m_{kr} of two random variables **x**, and **y** is defined as:

$$m_{kr} = E\{\mathbf{x}^k \mathbf{y}^r\} = \int_{-\infty}^{\infty}\int_{-\infty}^{\infty} x^k y^r f(x,y)\, dx dy, \tag{F.8}$$

Particular importance has the joint second-order moment m_{11} or *correlation*:

$$R_{xy} = E\{\mathbf{xy}\} = m_{11} \tag{F.9}$$

The generic $(k+r)^{th}$-order joint *central* moment μ_{kr} of two random variables is defined as:

$$\mu_{kr} = E\{(\mathbf{x} - \eta_x)^k (\mathbf{y} - \eta_x)^r\} =$$
$$= \int_{-\infty}^{\infty} \int_{-\infty}^{\infty} (\mathbf{x} - \eta_x)^k (\mathbf{y} - \eta_y)^r f(x,y) \, dxdy \quad \text{(F.10)}$$

The second central moment of two random variables is called the *covariance*:

$$\mu_{11} = E\{(\mathbf{x} - \eta_x)(\mathbf{y} - \eta_y)\} = \int_{-\infty}^{\infty} \int_{-\infty}^{\infty} (x - \eta_x)(y - \eta_y) f(x,y) \, dxdy \quad \text{(F.11)}$$

For the linearity of the expectation is:

$$\mu_{11} = E\{(\mathbf{x} - \eta_x)(\mathbf{y} - \eta_y)\} = E\{\mathbf{xy}\} - \eta_x E\{\mathbf{y}\} - \eta_y E\{\mathbf{x}\} + \eta_x \eta_y$$
$$= E\{\mathbf{xy}\} - E\{\mathbf{x}\}E\{\mathbf{y}\} \quad \text{(F.12)}$$

The ratio:

$$r = \frac{E\{(\mathbf{x} - \eta_x)(\mathbf{y} - \eta_y)\}}{\sqrt{E\{(\mathbf{x} - \eta_x)^2 (\mathbf{y} - \eta_y)^2\}}} = \frac{\mu_{11}}{\sigma_x \sigma_y} \quad \text{(F.13)}$$

is called *correlation coefficient* of **x** and **y**.

F.2.3 Uncorrelated, Orthogonal, Independent random variables

Two r. v. **x**, and **y** are called *uncorrelated* if

$$E\{\mathbf{xy}\} = E\{\mathbf{x}\}E\{\mathbf{y}\} \quad \text{(F.14)}$$

They are called *orthogonal* if

$$E\{\mathbf{xy}\} = 0 \quad \text{(F.15)}$$

They are called *independent* if

$$f(x,y) = f_{\mathbf{x}}(x)f_{\mathbf{y}}(y) \tag{F.16}$$

F.3 Stochastic Processes

"We are given an experiment $\mathfrak{E}$ specified by its outcomes ζ forming the space $\mathscr{S}$, by certain subsets of $\mathscr{S}$ called events, and by the probabilities of these events. To every outcome ζ we now assign, according to a certain rule, a time function $\mathbf{x}(t,\zeta)$ real or complex. We have thus created a family of functions, one for each ζ. This family is called a *stochastic process*." ([74]).

Fig. F.1 gives an example of a stochastic process by showing four possible *time realizations*, and two random variables *extracted* from the process at two given time instants. It can be seen that a single realization of the process corresponds to a time-signal, whose values at particular time instants are not predictable but depend on the specific realization. Those values, therefore, represent random variables extracted from the process at the chosen time instants. Therefore, considering the process at a given time point corresponds to defining a random variable $\mathbf{x}(t_i)$. Each defined r.v. is characterized by a distribution function $D[x(t_i)]$ and a probability density $f[x(t_i)]$, which are called the (first order) *statistics* of the process. The *order* of the statistics is given by the number of the considered time instants, and, therefore, by the number of the considered random variables. If the process is evaluated at the time points t_1, and t_2, the joint distribution $D[x(t_1), x(t_2)]$ and the probability density $f[x(t_1), x(t_2)]$ of the two associated random variables $\mathbf{x}(t_1)$, and $\mathbf{x}(t_2)$ form the second-order statistics of the stochastic process $\mathbf{x}(t)$.

Referring further to fig. F.1, two kind of averages are defined for a process: the *time averages*, which are "in the horizontal sense", and the *ensemble averages*, which are "in the vertical sense".

The time averages correspond to common characteristics of the deterministic signals, such as, for instance, mean value or average power. The ensemble averages are the moments of the relevant r.v. associated with the process and its statistics, like, for instance, expected value, variance, and autocorrelation. These quantities

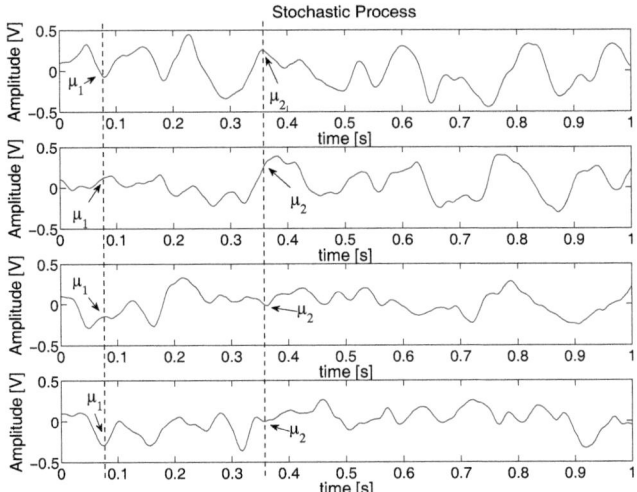

Figure F.1: *Some possible time realizations of a stochastic process* $\mathbf{x}(t,\zeta)$: *for a specific outcome* ζ_i, *the expression* $\mathbf{x}(t,\zeta_i)$ *represents a single time function. For a specific time* t_i, $\mathbf{x}(t_i,\zeta)$ *is a quantity depending on* ζ, *that is, a random variable* (μ_1 *and* μ_2 *in the graphic represents possible realizations of two such variables). Finally,* $\mathbf{x}(t_i,\zeta_i)$ *is a mere number.*([74])

are generally time-dependent, which makes the statistical analysis of a process a not affordable task, unless the considered process has the property of *stationarity*.

A stochastic process is said to be *strict sense stationary* (SSS) if its statistics are not affected by a shift in the time origin, that is, the two processes $\mathbf{x}(t)$ and $\mathbf{x}(t+\epsilon)$ have the same statistics for any ϵ. This implies, among other relationships,

$$E\{\mathbf{x(t)}\} = \eta_x = constant\,, \tag{F.17}$$

and

$$R(\tau) = E\{\mathbf{x(t}+\tau)\mathbf{x(t)}\} = R(-\tau)\,, \tag{F.18}$$

where $\tau = t_1 - t_2$.

If the stationarity holds only for the first two order statistics, the process is said to be *wide-sense* stationary (WSS). In this case, the relationships (F.17), and (F.18) still hold, so that in most cases wide-sense stationarity is sufficient to guarantee the possibility of analysis. Of course, if a process is SSS, it is also WSS.[2]

In analogy with the case of two r.v. (see eq.(F.12)), the *autocovariance* of a stochastic process is defined as:

$$C(t_1, t_2) = R(t_1, t_2) - \eta(t_1)\eta(t_2), \qquad (\text{F.19})$$

which, for WSS becomes:

$$C(\tau) = R(\tau) - \eta^2 \qquad (\text{F.20})$$

It must be stressed that the autocovariance $C(t_1, t_2)$ of a process is different from the variance of the r.v. $\mathbf{x}(t)$ extracted from the process:

$$C(t,t) = R(t,t) - \eta^2(t), \qquad (\text{F.21})$$

which, in case of a WSS process becomes:

$$C(0) = R(0) - \eta^2, \qquad (\text{F.22})$$

Furthermore, the power spectrum of a stationary process is defined as the Fourier transform of its autocorrelation:

$$S(f) = \mathscr{F}\{R(\tau)\} \qquad (\text{F.23})$$

---

[2] For Gaussian (or normal) processes, wide-sense stationarity implies strict-sense stationarity, because the statistics of a normal process are uniquely determined by its mean and autocorrelation.

F.3.1 Time averages, ensemble averages and ergodicity

Given a process $\mathbf{x}(t)$, we consider the two time averages:

$$\mathfrak{n} = \lim_{T \to \infty} \frac{1}{2T} \int_{-T}^{T} \mathbf{x}(t)\, dt, \qquad (F.24)$$

$$R(\tau) = \lim_{T \to \infty} \frac{1}{2T} \int_{-T}^{T} \mathbf{x}(t+\tau)\mathbf{x}(t)\, dt, \qquad (F.25)$$

which are called the *mean* and the *autocorrelation* of the process. This quantities are not constants, but random variables themselves. In order to claim that:

$$\mathfrak{n} = E\{\mathbf{x}(t)\} \quad \text{and} \quad R(\tau) = E\{\mathbf{x}(t+\tau)\mathbf{x}(t)\} \qquad (F.26)$$

it must be proven that:

$$E\{\mathfrak{n}\} = E\{\mathbf{x}(t)\} = \eta \qquad \text{with} \quad \sigma_\mathfrak{n}^2 = 0, \qquad (F.27)$$

and:

$$E\{R(\tau)\} = E\{\mathbf{x}(t+\tau)\mathbf{x}(t)\} = R(\tau) \qquad \text{with} \quad \sigma_R^2 = 0 \qquad (F.28)$$

Ergodicity deals with the problem of determining the statistics of a process $\mathbf{x}(t)$ from a single observation. There are two equivalent definitions of ergodicity:

1. $\mathbf{x}(t)$ is ergodic in the most general form if (with probability 1) all its statistics can be determined from a single function $\mathbf{x}(t, \zeta)$ of the process;

2. $\mathbf{x}(t)$ is ergodic if *time averages equal ensemble averages*;

Ergodicity its a fundamental characteristic for the analysis of a process, since it implies that a process can be characterized without knowing all its possible realizations.

Important relationships, which follow from ergodicity are shown by looking at the covariance in different cases:

(a) two random variables **x**, **y**: the covariance corresponds to the second central moment (see eq.(F.11)):

$$\mu_{11} = E\{(\mathbf{x} - \eta_x)(\mathbf{y} - \eta_y)\} = E\{\mathbf{xy}\} - E\{\mathbf{x}\}E\{\mathbf{y}\}; \tag{F.29}$$

(b) a stochastic process **x**(t): the covariance of two r.v. extracted at the time instants t_1, t_2 is called autocovariance of the process and is given by (see eq.(F.19)):

$$C(t_1, t_2) = R(t_1, t_2) - \eta(t_1)\eta(t_2); \tag{F.30}$$

(c) a stationarity stochastic process **x**(t) (see eq.(F.20)):

$$C(t_1, t_2) = C(\tau) = R(\tau) - \eta^2 \quad (\tau = t_1 - t_2); \tag{F.31}$$

which, for $\tau = 0$ becomes:

$$C(0) = R(0) - \eta^2. \tag{F.32}$$

If the analyzed process is also ergodic, the statistic autocorrelation coincides with the signal autocorrelation, and the statistic expected value coincides with the mean value of the signal, so that the relationship (F.32) can be interpreted in terms of *signal power*: since the autocorrelation of a signal evaluated at zero corresponds to its total average power (see eq. (6.38)), from (F.32) follows that the autocovariance of the process calculated at zero corresponds to the average AC power, being the difference of the total power and the DC power.

Appendix G

The Wigner Distribution[1]

All methods that have been used extensively for time-frequency analysis of signals are based on the assumption that on a short-time basis the signals are stationary. This has the important drawback that the length of the assumed short-time stationarity determines the frequency resolution which can be obtained. To increase the frequency resolution one has to take a longer measurement interval (window), which means that non-stationarities occurring during this interval will be smeared out in time and frequency.

A time-frequency characterization of a signal that overcomes this drawback is the *Wigner Distribution* (WD), which was introduced in 1932 by Wigner in the context of quantum mechanics ([103]). This operator is also known as Wigner-Ville Distribution, or *Wigner-Ville Transform*, as it was later re-derived by Ville in 1948 as a quadratic (in signal) representation of the local time-frequency energy of a signal ([102]).

This signal transformation has some important properties that make it an ideal tool for time-frequency signal analysis, as it allows to determine the value of the *instantaneous frequency* of a signal at each given time instant.

Although the concept as such was not new, it still was little known in the area of signal processing until 1980, when the WD obtained considerable attention in optics. Newly, the WD was introduced in Radio Occultation experiments, for the resolution of the ambiguity arising in the analysis of multipath-affected data ([35]).

In this Appendix a brief overview of the Wigner Distribution is given along with some of its properties which are relevant for signal processing, in particular for the

---
[1] An extensive treatment of this subject can be found in [12], [13], [14], and [15].

processing of multipath-affected signals from radio occultation measurements.

First, the WD is defined for analog signals (continuous-time signals) and afterwards adapted to the case of digital signals (discrete-time signals).

G.1 Analog signals

Given two continuous-time signals $f(t)$ and $g(t)$, where $f, g \in \mathbb{C}$, and $t \in \mathbb{R}$, the Wigner-Ville Transform, or cross-Wigner distribution $W_{f,g}$ of these functions is defined as:

$$W_{f,g}(t,\omega) = \int_{-\infty}^{\infty} e^{-j\omega\tau} f\left(t + \frac{\tau}{2}\right) g^*\left(t - \frac{\tau}{2}\right) d\tau \tag{G.1}$$

with $\omega \in \mathbb{R}$. The asterisk denotes the complex conjugated quantity. It is possible to show that the function $W_{f,g}$ is a real quantity ([15]) which depends on both time t and angular frequency ω, so that it is possible to associate a frequency value to each time instant and vice versa.

The auto-Wigner distribution of a signal $f(t)$ is:

$$W_{f,f}(t,\omega) = W_f(t,\omega) = \int_{-\infty}^{\infty} e^{-j\omega\tau} f\left(t + \frac{\tau}{2}\right) f^*\left(t - \frac{\tau}{2}\right) d\tau , \tag{G.2}$$

that is, the distribution $W_f(t,\omega)$ is capable of rendering the spectral contents of the signal $f(t)$ at each given time instant t_i with a frequency resolution which is independent of the time.

Indicating with $F(\omega)$ and $G(\omega)$ the Fourier spectra of $f(t)$ and $g(t)$, respectively, by defining the WD for the spectra as:

$$W_{F,G}(\omega, t) = \int_{-\infty}^{\infty} e^{j\omega\tau} F\left(\omega + \frac{\xi}{2}\right) G^*\left(\omega - \frac{\xi}{2}\right) d\tau; \tag{G.3}$$

one obtains the important relation:

$$W_{F,G}(\omega, t) = W_{f,g}(t, \omega); \tag{G.4}$$

Another important property of the WD can be derived by noting that eq. (G.1) can be interpreted as the Fourier Transform of the signal $f\left(t+\frac{\tau}{2}\right)g^*\left(t-\frac{\tau}{2}\right)$ in the variable τ, having t as a fixed parameter. The inverse Fourier transform can be expressed as:

$$\frac{1}{2\pi}\int_{-\infty}^{\infty} e^{j\omega\tau} W_{f,g}(t,\omega)\, d\omega = f\left(t+\frac{\tau}{2}\right) g^*\left(t-\frac{\tau}{2}\right) \qquad (G.5)$$

Letting

$$\begin{cases} t+\frac{\tau}{2} = t_1 \\ t-\frac{\tau}{2} = t_2 \end{cases} \qquad (G.6)$$

one obtains:

$$\frac{1}{2\pi}\int_{-\infty}^{\infty} e^{j\omega(t_1-t_2)} W_{f,g}\left(\frac{t_1+t_2}{2},\omega\right)\, d\omega = f(t_1)\, g^*(t_2); \qquad (G.7)$$

In the special case $t = t_1 = t_2$, the (G.6) becomes

$$\frac{1}{2\pi}\int_{-\infty}^{\infty} W_{f,g}(t,\omega)\, d\omega = f(t)\, g^*(t); \qquad (G.8)$$

in particular, for $g(t) = f(t)$,

$$\frac{1}{2\pi}\int_{-\infty}^{\infty} W_f(t,\omega)\, d\omega = |f(t)|^2. \qquad (G.9)$$

This means that the integral of the WD over the frequency variable at a certain time t yields the instantaneous signal power at that time. Integration of eq. (G.9) w.r.t. time yields:

$$\frac{1}{2\pi}\int_{t_a}^{t_b}\left[\int_{-\infty}^{\infty} W_f(t,\omega)\, d\omega\right] dt = \int_{t_a}^{t_b} |f(t)|^2\, dt. \qquad (G.10)$$

This relationships shows that the integral of the WD of the signal f over the infinite strip $-\infty < \omega < \infty$, $t_a < t < t_b$ is equal to the energy contained in $f(t)$ in the time interval $t_a < t < t_b$. The total energy in f is therefore given by the integral of the WD over the whole plane (t,ω):

$$\frac{1}{2\pi} \int_{-\infty}^{\infty} \int_{-\infty}^{\infty} W_f(t,\omega) \, d\omega \, dt = \int_{-\infty}^{\infty} |f(t)|^2 \, dt = (f,f) = ||f(t)||^2, \qquad \text{(G.11)}$$

where $(,)$ denotes *inner product*.[2]

Similarly to (G.5), noticing from (G.3) that $W_{F,G}(\omega,t)$ is the inverse Fourier transform of $F(\omega+\xi/2) \cdot G^*(\omega-\xi/2)$, considered as a function of ξ, having ω as a fixed parameter, one obtains:

$$\int_{-\infty}^{\infty} e^{-j\xi t} W_{F,G}(\omega,t) \, dt = F\left(\omega+\frac{\xi}{2}\right) G^*\left(\omega-\frac{\xi}{2}\right) \qquad \text{(G.12)}$$

Analogously to the argumentation expressed by equations (G.5) through (G.9), starting from eq. (G.12) and making use of (G.4), in the case $G(\omega) = F(\omega)$ one obtains:

$$\int_{-\infty}^{\infty} W_f(t,\omega) \, dt = |F(\omega)|^2; \qquad \text{(G.13)}$$

This means that the integral of the WD over time at certain frequency ω yields the energy density spectrum of f at this frequency. Integration of eq. (G.13) w.r.t. frequency yields:

$$\frac{1}{2\pi} \int_{\omega_a}^{\omega_b} \left[\int_{-\infty}^{\infty} W_f(t,\omega) \, dt \right] d\omega = \frac{1}{2\pi} \int_{\omega_a}^{\omega_b} |F(\omega)|^2 \, d\omega. \qquad \text{(G.14)}$$

This relationships shows that the integral of the WD of the signal f over the infinite strip $-\infty < t < \infty$, $\omega_a < \omega < \omega_b$ is equal to the energy contained in f in

---

[2]The inner product $(,)$ is defined in the vector space **V** of the time signals. The WD is considered as a *bilinear map* $\mathbf{V} \times \mathbf{V} \to \mathbf{F}$, where **F** is the scalar field $\mathbb{R}$, (see par.7.2.3).

the frequency interval $\omega_a < \omega < \omega_b$. This relationship is complementary to (G.10). Taking $\omega_a = -\infty$ and $\omega_b = \infty$ results again in eq. (G.11).

In order to characterize the WD more specifically, central moments can be defined. In particular, the first-order moment w.r.t. the frequency variable is:

$$\Omega_f(t) = \frac{\frac{1}{2\pi}\int_{-\infty}^{\infty} \omega W_f(t,\omega)\, d\omega}{\frac{1}{2\pi}\int_{-\infty}^{\infty} W_f(t,\omega)\, d\omega}; \tag{G.15}$$

which can be considered as the average frequency of the WD at time t; by recalling eq. (G.9), eq. (G.15) becomes

$$\Omega_f(t) = \frac{\frac{1}{2\pi}\int_{-\infty}^{\infty} \omega W_f(t,\omega)\, d\omega}{|f(t)|^2}. \tag{G.16}$$

It is possible to show that

$$\Omega_f(t) = \Im\mathfrak{m}\left\{\frac{f'(t)}{f(t)}\right\} = \Im\mathfrak{m}\left\{\frac{d}{dt}\left[\ln f(t)\right]\right\}. \tag{G.17}$$

Representing the signal $f(t)$ in the usual complex form

$$f(t) = v(t)e^{j\phi(t)} \tag{G.18}$$

where $v(t)$ is the time-varying amplitude of the signal and $\phi(t)$ is time-varying phase of the signal, it follows that:

$$\Omega_f(t) = \phi'(t). \tag{G.19}$$

Eq. (G.19) means that the average frequency of the WD at time t is equal to the derivative of the signal phase $\phi(t)$, that is, its *instantaneous frequency*.

Global moments of the WD are obtained by integration over the whole plane.

They are therefore constants, which characterize the WD in a global sense.

The average of the WD is (see eq. (G.11)):

$$\bar{P}_f = \frac{1}{2\pi} \int_{-\infty}^{\infty} \int_{-\infty}^{\infty} W_f(t,\omega) \, dt \, d\omega = \|f\|^2 = \|F\|^2 \qquad (G.20)$$

$\bar{P}_f$ is the total energy of the signal f, which is positive for all $f \neq 0$.

G.2 Discrete signals

In this section discrete-time signals, or *sequences*, of the type $f[n]$ are considered, where $f \in \mathbb{C}$, and $n \in \mathbb{Z}$. For these signals the Fourier transform is defined as:

$$F(\omega) = \sum_{n=-\infty}^{\infty} f[n] \, e^{-jn\omega}; \qquad (G.21)$$

with $\omega \in \mathbb{R}$. The inverse transform is given by:

$$f[n] = \frac{1}{2\pi} \int_{-\infty}^{\infty} F(\omega) e^{jn\omega} \, d\omega; \qquad (G.22)$$

The cross-Wigner distribution of two sequences $f[n]$ and $g[n]$ is defined as:

$$W_{f,g}(n,\omega) = 2 \sum_{k=-\infty}^{\infty} e^{-j2k\omega} f[n+k] g^*[n-k]; \qquad (G.23)$$

the auto-Wigner distribution of a sequence $f[n]$ is then given by:

$$W_f(n,\omega) = 2 \sum_{k=-\infty}^{\infty} e^{-j2k\omega} f[n+k] f^*[n-k]; \qquad (G.24)$$

the WD is a function of the discrete variable n and of the continuous variable ω; w.r.t. the latter variable the WD is periodic with period π (differently from the period 2π of all spectra of discrete-time signals):

$$W_{f,g}(n,\omega) = W_{f,g}(n,\omega+\pi), \quad \forall n,\omega. \tag{G.25}$$

In analogy with (G.4), also for discrete-time signals is:

$$W_{F,G}(\omega,t) = W_{f,g}(t,\omega); \tag{G.26}$$

furthermore,(see (G.9)):

$$\frac{1}{2\pi}\int_{-\pi/2}^{\pi/2} W_f(n,\omega)\,d\omega = |f[n]|^2; \tag{G.27}$$

and (see (G.13)):

$$\sum_{n=-\infty}^{\infty} W_f(n,\omega) = |F(\omega)|^2 + |F(\omega+\pi)|^2; \tag{G.28}$$

The relationship (G.28) shows that the energy density spectrum of the function f obtained by means of the WD is periodic with period π. This, generally, causes aliasing, i.e., frequency components that are π apart have the same influence on the WD.

There are two important situations, however, where the aliasing does not occur, because in these situations the spectrum $F(\omega)$ occupies only an interval of length π and is zero in the remaining part of one period of the spectrum. The first case is that of an "oversampled" signal, i.e., a signal with a band-limitation to less than $\pi/2$. Such a signal can be obtained either from an analogue signal by sampling it with a sampling frequency that is larger than twice the Nyquist rate or by interpolation of a discrete signal by a factor 2.

A second situation where only one of the terms on the right hand side in eq. (G.28) differs from zero arises when considering analytic signals, the spectrum of which vanishes for negative values of ω over the period of the spectrum. This is the case for the Open Loop data processing, since the input data are represented in complex form.

For the derivation of the moments in the discrete-time case it is referred to the indicated literature.

Appendix H

Quantization Error

We define a *quantization scale* by dividing the (expected) signal excursion Δ (absolute value of the difference between maximum- and minimum value of the signal amplitude) into N sub-intervals of uniform width $q = \Delta/N$, which we call *quantization intervals*. We then assign to each quantization interval a value $q_j = (\xi_{j+1}+\xi_j)/2$, where ξ_{j+1} and ξ_j are the values of the extremes of the j-th interval. The quantization process maps each value x of the sampled signal, (that is, each value of the random variable (r.v.) $\mathbf{x}$ extracted from the stochastic process $\mathbf{X}(t)$), to the value q_j which lies next to x on the quantization scale. In doing so, a *quantization error* $e = x - q_j$ arises, which is obviously a r.v. with zero expectation. In order to calculate the quantization error variance σ_e^2, we are firstly interested in the calculation of $\sigma_{e/\mathbf{x}\in I_j}^2$, which is the variance of the quantization error conditioned to the *event* $\{\mathbf{x} \in I_j\}$, where I_j is the j-th quantization interval. It holds ([17]):

$$\sigma_{e/\mathbf{x}\in I_j}^2 = \int_{\xi_j}^{\xi_{j+1}} (x - q_j)^2 \, p_\mathbf{x}(x/\mathbf{x} \in I_j) \, dx =$$

$$= \frac{1}{P\{\mathbf{x} \in I_j\}} \int_{\xi_j}^{\xi_{j+1}} (x - q_j)^2 \, p_\mathbf{x}(x) \, dx \qquad \text{(H.1)}$$

Since, by assuming small quantization intervals, the r.v. $\mathbf{x}$ can be considered to be *uniformly* distributed in the interval I_j, with $j \neq 1, N^1$, eq. (H.1) becomes ([17]):

$$\sigma_{e/\text{x} \in I_j}^2 = \frac{1}{\xi_{j+1} - \xi_j} \int_{\xi_j}^{\xi_{j+1}} (x - q_j)^2 \, dx = \frac{1}{12} (\xi_{j+1} - \xi_j)^2 = \frac{q^2}{12} \qquad (\text{H.2})$$

Considering that $E\{e\} = 0$, the variance of the quantization error is:

$$\sigma_e^2 = \int_{-\infty}^{\infty} e^2 \, p_e(e) \, de = \int_{-\infty}^{\infty} (x - q_j)^2 \, p_e(x - q_j) \, de =$$
$$= \int_{-\infty}^{\infty} (x - q_j)^2 \, p_{\text{x}}(x) \, dx \qquad (\text{H.3})$$

Since $p_{\text{x}}(x/\text{x} \in I_j) = \dfrac{p_{\text{x}}(x)}{P\{\text{x} \in I_j\}}$, we obtain:

$$\sigma_e^2 = \sum_{j=2}^{N-1} P\{\text{x} \in I_j\} \cdot \int_{\xi_j}^{\xi_{j+1}} (x - q_j)^2 \, p_{\text{x}}(x/\text{x} \in I_j) \, dx +$$
$$+ \int_{-\infty}^{\xi_2} (x - q_1)^2 \, p_{\text{x}}(x) \, dx + \int_{\xi_N}^{\infty} (x - q_N)^2 \, p_{\text{x}}(x) \, dx \qquad (\text{H.4})$$

The last two terms in eq. (H.4) accounts for the *saturation* error of the mapping $x \to q_j$, which is due to the fact that the signal to be digitized could exceed the expected excursion Δ. In this case, all signal values larger than $\Delta/2$ would be mapped into the same interval I_N, thus obtaining the same quantized value q_N, and, symmetrically, all signal values smaller than $-\Delta/2$ would be mapped into the same interval I_1, thus obtaining the same quantized value q_1.

Setting $\int_{-\infty}^{\xi_2} (x - q_1)^2 \, p_{\text{x}}(x) \, dx + \int_{\xi_N}^{\infty} (x - q_N)^2 \, p_{\text{x}}(x) \, dx = \sigma^2$, and recalling eq. (H.1), we obtain (see also [17]):

$$\sigma_e^2 = \sum_{j=2}^{N-1} P\{\text{x} \in I_j\} \cdot \sigma_{e/\text{x} \in I_j}^2 + \sigma^2 = \frac{q^2}{12} \cdot \int_{-\Delta/2}^{\Delta/2} p_{\text{x}}(x) \, dx + \sigma^2 \qquad (\text{H.5})$$

If we were to assume that the input signal does not exceed the extremes $\pm \Delta/2$

---

[1] For the extreme intervals the hypothesis of uniform *pdf* is not valid (see after).

of the quantization scale, eq. (H.4) would become:

$$\sigma_e^2 = \sum_{j=1}^{N} P\{\mathrm{x} \in I_j\} \cdot \int_{\xi_j}^{\xi_{j+1}} (x - q_j)^2 \, p_\mathrm{x}(x/\mathrm{x} \in I_j) \, dx = \frac{q^2}{12} \cdot \int_{-\Delta/2}^{\Delta/2} p_\mathrm{x}(x) \, dx \quad \text{(H.6)}$$

Since, due to the assumption made:

$$\int_{\Delta/2}^{\Delta/2} p_\mathrm{x}(x) \, dx = 1 \quad , \tag{H.7}$$

eq. (H.6) becomes:

$$\sigma_e^2 = \frac{q^2}{12} \cdot \int_{-\Delta/2}^{\Delta/2} p_\mathrm{x}(x) \, dx = \frac{q^2}{12} = \sigma_{e/\mathrm{x} \in I_j}^2 \quad , \tag{H.8}$$

that is, the variance σ_e^2 of the total quantization error coincide with the variance $\sigma_{e/\mathrm{x} \in I_j}^2$ of the error computed for the single quantization intervals. The quantity $q^2/12$, which was derived in this Appendix, is usually taken as reference for the variance of the quantization error, that is, for the *quantization noise power*.

Bibliography

[1] Abelló, R., *RSI IFMS SPECIFICATION*, GRST-NNO-IFMS-SP-1001-TOS-GSS, 15 Dec.2003.

[2] Agilent Technologies, *The Fundamentals of Signal Analysis*, Application Note 243. 2000.

[3] Andert, T.P., *Mass estimation of small solar system bodies using Radio Science data from close flybys*, PhD thesis, Universität der Bundeswehr München, Neubiberg, Germany, Fakultät fr Luft- und Raumfahrttechnik, Januar 2010.

[4] Baher, H., *Analog and Digital Signal Processing*. John Wiley & Sons Ltd., 1990.

[5] Barabash, S., Sauvaud, J.-A., Gunell, H., Andersson, H., Grigoriev, A., Brinkfeldt, K., Holmstrom, M., Lundin, R., Yamauchi, M., Asamura, K., Baumjohann, W., Zhang, T.L., Coates, A.J., Linder, D.R., Kataria, D.O., Curtis, C.C., Hsieh, K.C., Sandel, B.R., Fedorov, A., Mazelle, C., Thocaven, J.-J., Grande, M., Koskinen, H.E.J., Kallio, E., Sales, T., Riihela, P., Kozyra, J., Krupp, N., Woch, J., Luhmann, J., McKenna-Lawlor, S., Orsini, S., Cerulli-Irelli, R., Mura, M., Milillo, M., Maggi, M., Roelof, E., Brandt, P., Russell, C.T., Szego, K., Winningham, J.D., Frahm, R.A., Scherrer, J., Sharber, J.R., Wurz, P., Bochsler, P., 2006. *The analyzer of space plasmas and energetic atoms (ASPERA-4) for the Venus Express mission.* Planet. Space Sci. 55, 2007.

[6] Barriere, J., Benvenuto, J.-M., Clochet, A., Gillot, B., Midan, N., Penalva, S., Riant, P., Sibilla, C., Trillard, D., Schirmann, T., Osswald, Fabrega, J. *VENUS EXPRESS SPACECRAFT DESIGN REPORT 0*, 01-VEX.T.ASTR.TCN.00349, I2, R0, 06.02.2004.

[7] Bertaux, J.-L., Nevejans, D., Korablev, O., Villard, E., Quémerais, E., Neefs, E., Montmessin, F., Leblanc, F., Dubois, J.P., Dimarellis, E., Hauchecorne, A., Lefèvre, F., Rannou, P., Chaufray, J.Y., Cabane, M., Cernogora, G., Souchon, G., Semelin, F., Reberac, A., Van Ransbeek, E., Berkenbosch, S., Clairquin, R., Muller, C., Forget, F., Hourdin, F., Talagrand, O., Rodin, A., Fedorova, A., Stepanov, A., Vinogradov, I., Kiselev, A., Kalinnikov, Yu., Durry, G., Sandel, B., Stern, A., Gérard, J.C., 2006. *SPICAV/SOIR on Venus Express: three spectrometers to study the global structure and composition of the Venus atmosphere*. Planet. Space Sci. 55, 2007.

[8] Born, M., Wolf, E., *Principles of Optics*. Pergamon Press, Oxford, 1971, pp. 121127.

[9] Bracewell, R. N. *The Fourier Transform and Its Application*. Revision of second edition, 1986, McGraw-Hill Book Company.

[10] *"Der Brockhaus"*. *Astronomie.*, F.A. Brockhaus GmbH, Leipzig - Mannheim, 2006.

[11] Campbell, B. A., *Radar Remote Sensing of Planetary Surfaces*. Cambridge University Press, 2002.

[12] Claasen, T.A.C.M., Mecklenbruker, W.F.G. *The Wigner Distribution - A tool for time-frequency signal analysis. Part I: Continuous-time signals*, Philips J. Res. 35, 217-250, 1980.

[13] Claasen, T.A.C.M., Mecklenbruker, W.F.G. *The Wigner Distribution - A tool for time-frequency signal analysis. Part II: Discrete-time signals*, Philips J. Res. 35, 276-300, 1980.

[14] Claasen, T.A.C.M., Mecklenbruker, W.F.G. *The Wigner Distribution - A tool for time-frequency signal analysis. Part III: Relations with other time-frequency signal transformations.*, Philips J. Res. 35, 372-389, 1980.

[15] Claasen, T.A.C.M., Mecklenbruker, W.F.G. *Time-Frequency signal analysis by means of the Wigner Distribution.*, Acoustics, Speech, and Signal Processing, IEEE International Conference on ICASSP '81, Vol.6, 69-72, 1981.

[16] De Tiberis, F., *VEX FM Test Report*, RPT-VEX-0052-ALS, Alenia Spazio S.p.A., May 13, 2004.

[17] DiBenedetto, M.G., Mandarini, P. *Comunicazioni elettriche*, Edizioni Ingegneria 2000, 1999.

[18] Drossart, P., Piccioni, G., Adriani, A., Angrilli, F., Arnold, G., Baines, K.H., Bellucci, G., Benkhoff, J., Bezard, B., Bibring, J.-P., Blanco, A., Blecka, M.I., Carlson, R.W., Coradini, A., Di Lellis, A., Encrenaz, Th., Erard, S., Fonti, S., Formisano, V., Fouchet, T., Garcia, R., Haus, R., Helbert, J., Ignatiev, N.I., Irwin, P.G.J., Langevin, Y., Lebonnois, S., Lopez-Valverde, M.A., Luz, D., Marinangeli, L., Orofino, V., Rodin, A.V., Roos-Serote, M.C., Saggin, B., Sanchez- Lavega, A., Stam, D.M., Taylor, F.W., Titov, D.V., Visconti, G., Zambelli, M., Hueso, R., Tsang, C.C.C., Wilson, W.F., Afanasenko, T.Z., 2006. *Scientific goals for the observations of Venus by VIRTIS on ESA/ Venus Express mission*. Planet. Space Sci. 55, 2007.

[19] EADS Astrium, *VENUS EXPRESS SPACECRAFT USER MANUAL. Volume 3: Spacecraft Operations*, VEX-T.ASTR-UM-01098, I4.1, June 16, 2006.

[20] ESOC, European Space Operation Centre: *Mission Control System (MCS) Data Delivery Interface Document DDID*, RO-ESC-IF-5003/MEX-ESC-IF-5003/VEX-ESC-IF-5003, Appendix H: FD Products, I3.1, July 1, 2005.

[21] ESA, European Space Agency, *Science Homepage*, 2010. http://sci.esa.int/.

[22] ESA, European Space Agency, *Rosetta Homepage*, 2010. http://www.esa.int/SPECIALS/Rosetta/index.html

[23] Eshleman, V.R., *THE RADIO OCCULTATION METHOD FOR THE STUDY OF PLANETARY ATMOSPHERES*, Planet. Space Sci., Vol. 21, pp.1521-1531, 1973.

[24] Essen, L., and K. D. Froome (1951), *The refractive indices and dielectric constants of air and its principal constituents at $24,000 Mc/s$*, Proc. Phys. Soc. London, Sect. B, 64, 862 875, doi:10.1088/0370-1301/64/10/303.

[25] Formisano, V., Angrilli, F., Arnold, G., Atreya, S., Baines, K.H., Bellucci, G., Bezard, B., Billebaud, F., Biondi, D., Blecka, M.I., Colangeli, L., Comolli, L.,

Crisp, D., DAmore, M., Encrenaz, T., Ekonomov, A., Esposito, F., Fiorenza, C., Fonti, S., Giuranna, M., Grassi, D., Grieger, B., Grigoriev, J., Helbert, J., Hirsch, H., Ignatiev, N., Jurewicz, A., Khatuntsev, I., Lebonnois, S., Lellouch, E., Mattana, A., Maturilli, A., Mencarelli, E., Michalska, M., Lopez Moreno, J., Moshkin, B., Nespoli, F., Nikolsky, Yu., Nuccilli, F., Orleanski, P., Palomba, E., Piccioni, G., Rataj, M., Rinaldi, G., Rossi, M., Saggin, B., Stam, D., Titov, D., Visconti, G., Zasova, L., 2006. *The Planetary Fourier Spectrometer (PFS) onboard the European Venus Express mission.* Planet. Space Sci. 54, 2006.

[26] Fjeldbo, G., *Bistatic Radar Methods for Studying Planetary Ionspheres and Surfaces*, Scientific Report No.2, NsG-377, SU-SEL-64-025. Stanford Electronic Laboratories, Stanford University California, 1964.

[27] Fjeldbo, G., Kliore, A.J., Eshlemann, *The Neutral Atmosphere of Venus as Studied with the Mariner V Radio Occultation Experimens*, 1971. Astr. J. 76, 123.

[28] Glassmeier, K.H., Boehnhardt, H., Koschny, D., Khrt, E., Richter, I. *The Rosetta Mission: Flying Towards the Origin of the Solar System.* Published on: Schulz, R., Alexander, C., Boehnhardt, H., Glassmeier, K.H., *Rosetta*, Springer Science and Business Media, LLC, 2009.

[29] Goodman, J.W., *Introduction to Fourier Optics*, San Francisco, McGraw-Hill Book Company, 1968.

[30] Gorbunov, M.E., Gurvich, A.S., *Algorithms of inversion of Microlab-1 satellite data including effects of multipath propagation*, Int. J. Remote Sensing, Vol. 19, No. 12, 2283-2300, 1998.

[31] Gorbunov, M.E., Gurvich, A.S., *Microlab-1 experiment: Multipath effects in the lower troposphere*, J. Geophys. Res. 103, 1381913826, 1998.

[32] Gorbunov, M.E., *Radioholographic methods for processing radio occultation data in multipath regions*, Danish Meteorological Institute, Scientific Report, 2001.

[33] Gorbunov, M.E., *Radioholographic analysis of radio occultation data in multipath zones*, Radio Science, vol. 37, No. 1, 10.1029/2000RS002577, 2002.

[34] Gorbunov, M.E., *Radio-holographic analysis of Microlab-1 radio occultation data in the lower troposphere*, J. Geophys. Res. 107, NO. D12, 10.1029/2001JD000889, 2002.

[35] Gorbunov, M.E., Lauritsen, K.B., Leroy, S.S., *Application of Wigner distribution function for analysis of radio occultations*, to be published in Radio Science, 2010.

[36] Hagl, D., Remus, S., Häusler, B., *USO Phase Noise Performance of EM at $38\,MHz$ & FM at $5\,MHz$*, VEX-VERA-UBW-RP-4100, I1.0, 09.02.2004.

[37] Hamilton Technical Services, *Stable 32. Frequency Stability Analysis. User Manual. Version 1.2*, Hamilton Technical Services, 195 Woodbury Street. S. Hamilton, MA 01982 USA.

[38] Han, C.S., Tyler, G.L., *Characterizing and Resolving Diffractive Structures in Planetary Radio Occultation Using Multiple Phase Screens*, U.R.S.I., Maastricht, the Netherlands, 17-24 August 2002.

[39] Häusler, B., Zeiler, O., Billig, G., Eidel, W. *Satellitensimulator für Kleinsatelliten*, Bericht, Institut für Raumfahrttechnik, Universität der Bundeswehr Mnchen, 1996.

[40] Häusler, B., Eidel, W., Hagl, D., Remus, S., Selle, J., Pätzold, M., *Venus Express Radio Science Experiment VeRa, Reference Systems and Techniques used for the simulation and prediction of atmospheric and ionospheric sounding measurements at planet Venus*, Forschungsbericht LRT-WE-9-FB-4 (2003), Universität der Bundeswehr München, Germany.

[41] Häusler, B., Remus, S., Mattei, R., *USO FM Performance Test at $38\,MHz$*, VEX-VERA-UBW-RP-4200, I1.2, 12.07.2004.

[42] Häusler, B. and Remus, S., *VeRa Experiment Integrated Test Report: Flight Model Transponder & Engineering Model USO*, VEX-VERA-UBW-TR-4001, I1.0, 09.10.2004.

[43] Häusler, B., Pätzold, M., Mattei, R., Remus, S. *Venus Express Radio Science Experiment VeRa. Calibration Report*, VEX-VERA-UBW-AN-3005, I1.0, 17.06.2005.

[44] Häusler, B., Remus, S., Mattei, R., *Venus Express Radio Science Experiment VeRa. USO-TRSP Troubleshooting Test Report: Flight Model Transponder & Flight Model USO*, VEX-VERA-UBW-TR-4005, I2.0, 03.03.2005.

[45] Häusler, B., Pätzold, M., Tyler, G.L., Barriot, V.J.-P., Bird, M.K., Dehant, V., Hinson, D., Simpson, R.A., Treumann, R.A., Eidel, W., Mattei, R., Rosenblatt, P., Remus, S., Selle, J., Tellmann, S., 2006a. *Venus atmospheric, ionospheric, surface, and interplanetary radio wave propagation studies with the Venus Express radio science experiment VeRa.* ESA SP-1291, Noordwijk, The Netherlands, 2007.

[46] Häusler, B., Pätzold, M., Tyler, G.L., Simpson, R.A., Bird, M.K., Dehant, V., Barriot, J.-P., Eidel, W., Mattei, R., Remus, S., Selle, J., Tellmann, S., Imamura, T., 2006b. *Radio science investigations by VeRa onboard the Venus Express spacecraft.* Planet. Space Sci. 54, 2006.

[47] Häusler, B., Eidel, W., Mattei, R., Remus,Pätzold, M., Tellmann, S., *The Planetary Atmospheric Doppler Effect in a Relativistic Treatment*, Forschungsbericht LRT-9-FB-5 (2007), Universität der Bundeswehr München, Germany.

[48] Häusler, B., *Satellitensysteme.* Lecture Notes. 2007.

[49] Häusler, B., *private communication*, November 2010.

[50] Han, C.S., Tyler, G.L., 2003. *Resolving diffractive and guiding structures in thick atmospheres.* Geophys. Res. Abstr. 5, 14285.

[51] Hinson, D.P., Flasar, F.M., Kliore, A.J., Schinder, P.J., Twicken, J.D., Herrera, R.G., *Jupiters ionosphere: Results from the first Galileo radio occultation experiments.*, Geophys. Res. Lett., 24(17), 21072110, doi:10.1029/97GL01608, 1997.

[52] Hinson, D.P., Twicken, J.D., Karayel, E.T., 1998. *Jupiters ionosphere: new results from Voyager 2 radio occultation measurements.* J. Geophys. Res. 103, 95059520, 1998.

[53] Igarashi, K., Pavelyev, A., Hocke, K., Pavelyev, D., Kucherjavenkov, I.A., Matyugov, S., Zakharov, A., Yakovlev, O., *Radio holographic principle for*

observing natural processes in the atmosphere and retrieving meteorological parameters from radio occultation data, Earth Planets Space, 52, 893-899, 2000.

[54] James, N., *Radio Science IFMS WO 28/03 Test ReportWork Package 5000*, GSY/040243/104266, BAE SYSTEMS, 2004.

[55] James, N., *ICD for the IFMS ESU datasets*, GSY/050042/107853, BAE SYSTEMS, 2007.

[56] Jenkins, J.M., Steffes, P.G., Hinson, D.P., Twicken, J.D., Tyler, G.L., *Radio Occultation Studies of the Venus Atmosphere with the Magellan Spacecraft; 2. Results from the October 1991 Experiments*, ICARUS 110, 79-94, 1994.

[57] Karayel, E.T., Hinson, D.P., 1997. *Sub-Fresnel-scale vertical resolution in atmospheric profiles from radio occultation.*, Radio Science, Vol. 32, No. 2, 411423, 1997.

[58] Kay, S.M., *Modern Spectral Estimation. Theory & Application*. Prentice Hall, Englewood Cliffs, New Jersey, 1988.

[59] Kraus, J. D. *Antennas*. McGraw-Hill, 1998.

[60] Krüger, C. *Berechnung von möglichen Absorptionsschichten in der Venus-Atmosphäre aus Daten des Venus Express Radio Science Experimentes VeRa*. Studienarbeit II, Universität der Bundeswehr München, Neubiberg, Germany, Fakultät fr Luft- und Raumfahrttechnik, February 2008.

[61] Lamy, P.L., Toth, I., Davidsson, B.J.R., Groussin, O., Gutierrez, P., Jorda, L., Kaasalainen, M., Lowry, S.C., *The Nucleus of Comet 67P/Churyumov-Gerasimenko*. Published on: Schulz, R., Alexander, C., Boehnhardt, H., Glassmeier, K.H., *Rosetta*, Springer Science and Business Media, LLC, 2009.

[62] Lee, S.W., *Magellan Venus Radio Occultation Atmospheric Profiles Data Set Archive*, Volume MG_2401, Planetary Data System (PDS) Atmospheres (ATM) Discipline Node (DN), University of Colorado, http://atmos.nmsu.edu/pdsd/archive/data/mgn-v-rss-5-occ-prof-rtpd-v10/mg_2401/. 1996.

[63] Lindal, G.F., Lyons, J.R., Sweetnam, D.N., Eshleman, V.R., Hinson, D.P., Tyler, G.L., *The Atmosphere of Uranus: Results of Radio Occultation Measurements With Voyager 2*, J. Geophys. Res., 92, 14987-15001, 1987.

[64] Lipa, B., Tyler, G. L., *Statistical And Computational Uncertainties in Atmospheric Profiles From Radio Occultation: Mariner 10 At Venus*, Icarus 39, 192-208 (1979)

[65] Markiewicz, W.J., Titov, D.V., Ignatiev, N., Keller, H.U., Crisp, D., Limaye, S.S., Jaumann, R., Moissl, R., Thomas, N., Esposito, L., Watanabe, S., Fiethe, B., Behnke, T., Szemerey, I., Michalik, H., Perplies, H., Wedemeier, M., Sebastian, I., Boogaerts, W., Dierker, C., Osterloh, B., Bo ker, W., Koch, M., Michaelis, H., Belyaev, D., Dannenberg, A., Tschimmel, M., Russo, P., *Venus Monitoring Camera for Venus Express*. Planet. Space Sci. 55, 2007.

[66] Mattei, R., Remus, S., Häusler, B., *Venus Express Radio Science Experiment VeRa. Technical Note: Multiple reflections in a mismatched line*, VEX-VERA-UBW-TN-4020, I1.0, 21.10.2004.

[67] Mattei, R., Remus, S., Häusler, B., *VeRa Experiment Integrated Test Report: Flight Mode Transponder & Flight Model USO*, VEX-VERA-UBW-TR-4002, I1.1, 17.12.2004.

[68] Mattei, R., Häusler, B. , Pätzold, M., Remus, S., Eidel, W., Tellmann, S. , Andert, T., Selle, J., Bird, M. K. , Simpson, R. A. Tyler, G. L., Dehant, V., Asmar, S., Barriot, J.-P., Imamura, T., *The Radio Science Experiment "VeRa" Onboard Esa's Venus Express Spacecraft*, 1st CEAS, European Air and Space Conference, 10.-13. September 2007, Berlin, 2055-2064, CEAS-2007-121, Conference Proceedings.

[69] Melbourne, W., G. *Radio Occultation Using Earth Satellites: A Wave Theory Treatment*, Deep Space Communications and Navigation Series, JPL-CalTech, 2004.

[70] Mohr, P. J. , Taylor, B. N., Newell, D. B. *CODATA Recommended Values of the Fundamental Physical Constants: 2006*, National Institute of Standards and Technology, Gaithersburg, Maryland 20899-8401, USA.

[71] Oppenheim, A.V., Schafer, R.W., Buck, J.R., *Discrete-Time Signal Processing*, Prentice-Hall, Inc. 1989.

[72] Pätzold, M., Häusler, B., Bird, M.K., Tellmann, S., Mattei, R., Asmar, S.W., Dehant, V., Eidel, W., Imamura, T., Simpson, R.A. & Tyler, G.L., *The structure of Venus middle atmosphere and ionosphere*, Nature, 450, 657 660, doi:10.1038/nature06239.

[73] Pätzold, M., Häusler, B., Aksnes, K., Anderson, J.D., Asmar, S.W., Barriot, J.-P., Bird, M. K., Boehnhardt, H., Eidel, W., Grn, E., Ip, W.H., Marouf, E., Morley, T., Thomas, N., Tsurutani, B.T., Wallis, M.K., Mysen, E., Olson, O., Remus, S., Tellmann, S. , Andert, T., Carone, L., Fels, M., Stanzel, C. *Rosetta Radio Science Investigation (RSI)*. Published on: Schulz, R., Alexander, C., Boehnhardt, H., Glassmeier, K.H., *Rosetta*, Springer Science and Business Media, LLC, 2009.

[74] Papoulis, A. *Probability, Random Variables and Stochastic Processes*, McGraw-Hill, Inc, 1965.

[75] Pavelyev, A.G., *On the feasibility of radioholographic investigations of wave fields near the Earth's radio-shadow zone on the satellite-to-satellite path*, J. Commun. Technol. Electron., 43(8), 875-879, 1998.

[76] Pawlitzki, A. and Schwall, Th., *VeRa USO Detailed Design Document,* , VE-TIM-DD-3001, I5.0, 12.02.2004.

[77] Picardi, G., *Elaborazione del segnale radar*, Franco Angeli s.r.l., Milano, 1995.

[78] Piccioni, G, Drossart, P., Suetta, E., Cosi, M., Ammannito, E., Barbis, A., Berlin, R., Boccaccini, A., Bonello, G., Bouye, M., Capaccioni, F., Cherubini, G., Dami, M., Dupuis, O., Fave, A., Filacchione, G., Hello, Y., Henry, F., Hofer, S., Huntzinger, G., Melchiorri, R., Parisot, J., Pasqui, C., Peter, G., Pompei, C., Reess, J.M., Semery, A., Soufflot, A., Adriani, A., Angrilli, F., Arnold, G., Baines, K., Bellucci, G., Benkhoff, J., Bezard, B., Bibring, J.-P., Blanco, A., Blecka, M.I., Carlson, R., Coradini, A., Di Lellis, A., Encrenaz, T., Erard, S., Fonti, S., Formisano, V., Fouchet, T., Garcia, R., Haus, R., Helbert, J., Ignatiev, N.I., Irwin, P., Langevin, Y., Lebonnois, S., Lopez Valverde, M.A., Luz, D., Marinangeli, L., Orofino, V., Rodin, A., Roos-Serote, M.C.,

Saggin, B., Sanchez-Lavega, A., Stam, D., Taylor, F., Titov, D.V., Visconti, G., Zambelli, M., 2006. *VIRTIS (Visible and Infrared Thermal Imaging Spectrometer) for Venus Express*. ESA SP-1291, Noordwijk, The Netherlands, in press.

[79] Remus, S., *Untersuchungen Zur Durchführung Von Satellitengestützten Radio Science Experimenten Im Interplanetaren Raum*, PhD thesis, Universität der Bundeswehr München, Neubiberg, Germany, Fakultät für Luft- und Raumfahrttechnik, June 2004.

[80] Remus, S., *private communication*, July 2010.

[81] Remus, S., *private communication*, October 2010.

[82] Ricart, M., *IFMS SUM (Software User Manual*, MakaluMedia/MR/IFMS/SUM, I11.2.0, 2006.

[83] Rutman, J., *Characterization of Phase and Frequency Instabilities in Prescision Frequency Sources: Fifteen Years of Progress*, IEEE Proceedings, Vol. 66, No. 9, 1048-1075, September 1978.

[84] Schaa, R., *Abel-Inversion von Okkultationsdaten*, Diplomarbeit, Universität zu Köln, Germany, Fakultät fr Geophysic, Januar 2005.

[85] Seidelmann, P.K. and Fukushima,T., *Why new time scales?*, Astron.Astrophys.265, 833-838, 1992.

[86] Selle, J., *Planung und Simulationen von Radio-Science-Experimenten Interplanetarer Raumfahrt-missionen*, PhD thesis, Universität der Bundeswehr München, Neubiberg, Germany, Fakultät für Luft- und Raumfahrttechnik, December 2005.

[87] Simpson, R.A., *Spacecraft Studies of Planetary Surfaces using Bistatic Radar*, IEEE Transact. Geosci. Rem. Sens. 31,2,465,1993.

[88] Simpson, R. A., Tyler, G. L., Pätzold, M., and Häusler, B., *Determination of local surface properties using Mars Express bistatic radar*, J. Geophys. Res., 111, E06S05, doi:10.1029/2005JE002580, 2006.

[89] Simpson, R. A., Tyler, G. L., Häusler, B., and Pätzold, M., (2007), *Venus Express bistatic radar experiments at Maxwell Montes*, EGU 2007-A-10326, European Geosciences Union (EGU), General Assembly, Vienna, 16.-20.4.2007.

[90] Simpson, R. A., Tyler, G. L., Häusler, B., Mattei, R., and Pätzold, M., *Venus Express bistatic radar: High-elevation anomalous reflectivity*, J. Geophys. Res., 114, E00B41, doi:10.1029/2008JE003156, 2009.

[91] Sokolovskiy, S.V., *Modelling and inverting radio occultation signals in the moist troposphere*, Radio Science, vol. 36, issue 3, 441-458, 2001.

[92] Stewart, R.W., Hogan, J.S, *Error analysis for the Mariner-6 and -7 occultation experiments*,Radio Sci. 8, 2, 109115. 1973.

[93] Svedhem, H., Titov, D.V., McCoy, D., Lebreton, J.P., Barabash, S., Bertaux, J.-L., Drossart, P., Formisano, V., Häusler, B., Korablev, O., Markiewicz, W.J., Nevejans, D., Pätzold, M., Piccioni, G., Zhang, T.L., Taylor, F.W., Lellouch, E., Koschny, D., Witasse, O., Warhaut, M., Accomazzo, A., Rodriguez-Canabal, J., Fabrega, J., Schirrmann, T., Clochet, A., Coradini, M., *Venus Express - The first European mission to Venus*, Planet. Space Sci. 55, 1636-1652, 2007.

[94] Tellmann, S., Pätzold, M., Häusler, B., Bird, M.K., and Tyler, G.L. *Structure of the Venus neutral atmosphere as observed by the Radio Science experiment VeRa on Venus Express.*, J.Geophys. Res. 114, E00B36, doi:10.1029/2008JE003204, 2009.

[95] Tellmann, S., *private communication*, July 2010.

[96] Titov, D.V., Svedhem, H., Koschny, D., Hoofs, R., Barabash, S., Bertaux, J.-L., Drossart, P., Formisano, V., Husler, B., Korablev, O., Markiewicz, W.J., Vevejans, D., Ptzold, M., Piccioni, G., Zhang, T.L., Merritt, D., Witasse, O., Zender, J., Accomazzo, A., Sweeney, M., Trillard, D., Janvier, M., Clochet, A., *Venus Express science planning*, Planet. Space Sci. 54, 1279-1297, 2006.

[97] Tyler, G.L., Ingalls, D.H.H., *Functional Dependences of Bistatic-Radar Frequency Spectra and Cross Sections on Surface Scattering Laws*, J.Geophys. Res. 76, 1971.

[98] Tyler, G.L., *private communication*, March 2010.

[99] Ulaby, F. T., Moore, R. K., Fung, A. K. *Microwave remote sensing active and passive*, VOL I. Addison-Wesley Publishing Company, 1981.

[100] Venus Express Ground Segment Team. *Space/Ground Interface Control Document*, VEX-ESC-IF-5002, I2.4., 11 July 2005.

[101] Villard, O.G., *The ionospheric sounder and its place in the history of radio science*, Radio Science, vol. 11, issue 11, 847-860, 1976.

[102] Ville, J., *Theory and applications of the notion of complex signal*, The Rand Corporation, translated from the French by I. Selin, August 1, 1958.

[103] Wigner, E., *On the Quantum Correction For Thermodynamic Equilibrium*, Phys. Rev., 749-759, 1932.

[104] Yuen, J.H., *Deep Space Telecommunication Systems Engineering*, Plenum Press, 1983.

[105] Zhang, T.L., Baumjohann, W., Delva, M., Auster, H.-U., Balogh, A., Russell, C.T., Barabash, S., Balikhin, M., Berghofer, G., Biernat, H.K., Lammer, H., Lichtenegger, H., Magnes, W., Nakamura, R., Penz, T., Schwingenschuh, K., Vörös, Z., Zambelli, W., Fornacon, K.-H., Glassmeier, K.-H., Richter, I., Carr, C., Kudela, K., Shi, J.K., Zhao, H., Motschmann, U., Lebreton, J.-P., *Magnetic field investigation of the Venus plasma environment: expected new results*. Planet. Space Sci. 54, 2006.

Die VDM Verlagsservicegesellschaft sucht für wissenschaftliche Verlage abgeschlossene und herausragende

Dissertationen, Habilitationen, Diplomarbeiten, Master Theses, Magisterarbeiten usw.

für die kostenlose Publikation als Fachbuch.

Sie verfügen über eine Arbeit, die hohen inhaltlichen und formalen Ansprüchen genügt, und haben Interesse an einer honorarvergüteten Publikation?

Dann senden Sie bitte erste Informationen über sich und Ihre Arbeit per Email an *info@vdm-vsg.de*.

Sie erhalten kurzfristig unser Feedback!

VDM Verlagsservicegesellschaft mbH
Dudweiler Landstr. 99   Telefon +49 681 3720 174
D - 66123 Saarbrücken   Fax   +49 681 3720 1749
www.vdm-vsg.de

Die VDM Verlagsservicegesellschaft mbH vertritt

Printed by Books on Demand GmbH, Norderstedt / Germany